Structural Periodic Table

From the Chemical Periodic Table to the Origin of Life

John Chang

(Hai Zhi Tao)

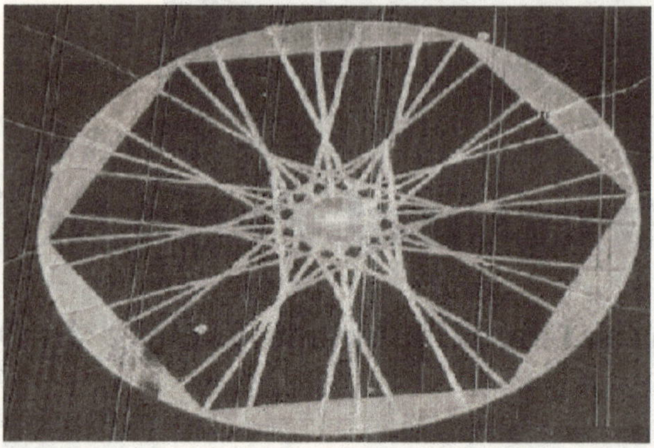

Every major scientific revolution has introduced a new language for understanding nature. Newton described nature through motion. Einstein described nature through geometry. The Structural Periodic Table explores whether structure itself may serve as a universal language connecting different scientific disciplines. It does not seek to replace existing theories, but to complement them by explaining why so many of them exhibit common structural patterns.

Publishing:

July, 2026 1st edit English ISBN: **978-1-7643097-6-9**

Universal Publishing

Author: John Chang

The author, known by the pen name *Hai Zhi Tao*, is also identified as Zhang Tao (Chinese) and John Chang (English). He was born in January 1961 in Beijing, China. He studied at Beijing No. 51 Middle School, the Wuhan Air Force Radar Institute, Beijing Textile Engineering Institute, TAFE Sydney, and the University of Technology Sydney (UTS).

Due to strong dissatisfaction with traditional education systems characterized by rote learning and credential-oriented training, he withdrew from multiple institutions and pursued independent study. Over the years, he developed deep interests in philosophy, physics, astronomy, religion, and economics, with particular attention to foundational questions about knowledge, structure, and reality.

After moving to Australia in 1990, he began exploring religious systems in depth. While initially unfamiliar with religion, he studied various traditions and participated in church activities.

He also studied economics, particularly stock market behavior, developing a strong interest in financial systems. and collective human dynamics.

A turning point came on September 11, 2001, when the attacks on the World Trade Center prompted him to deeply reflect on human society, religion, science, and the nature of reality. Shortly thereafter, during a fishing trip by the sea, he observed raindrops striking the water surface and generating expanding waves — an insight that led to what he describes as the realization of a fundamental “universal law.”

He spent two years writing his first major work, *Universal Law*, published in November 2003. Including the years of prior conceptual accumulation, the development of this work spanned nearly two decades. The book represents an attempt to unify scientific and philosophical understanding within a single structural framework.

Following this, he established several institutions dedicated to promoting his theoretical system worldwide, including: Universal Law Research Society; Universal Law Academy; Universal Publishing; Universal Law Art Institute.

His work has since expanded into a large-scale theoretical project aimed at unifying mathematics, physics, chemistry, and broader systems of knowledge under a structural framework based on recursive and generative principles.

Selected Publications

- *Universal Law* (2003.11)
- *Golden Classic* (2007.1)
- *Chinese Systems Philosophy* (2010.1, author & editor)
- *Oriental Systems Literature* (2010,1, author & editor)
- *Grand Ultimate Theory* (2013.9)
- *Crop Circle* (2015.6)
- *World Systems Science* (2018.4 author & editor)
- *Memoirs of Beijing No. 51 Middle School* (2024.3 author & editor)
- *Eastern Galactic Civilization: Unifying Religion, Philosophy and Science* (2024.7)
- *The Ultra Grand Unified Structure: From the Generative Law (•, 1, O) to the Recursive Equation of Everything* (2025.4)
- *Recursive Self-Organizing Evolutionary Breakthrough: Unifying Natural, Social and Life Sciences* (2025.8)
- *Life Intelligence Wave: Unifying Mathematics, Physics and Chemistry* (2025.12)
- *Metastructural Unification (Mathematics Volume): Unifying Algebra, Geometry and Probability* (2026.2)
- *Metastructural Unification (Physics Volume): Unifying Electromagnetism, Gravitation and Quantum Theory* (2026.4)
- *Metastructural Unification (Chemistry Volume): Unifying Reactions, Compounds and Elements (*2026.5*)*

Website: http://www.universal-law68.com

Preface

Structural Periodic Table ----- From the Chemical Periodic Table to the Origin of Life

---

I. The Beginning of the Question

The Chemical Periodic Table was humanity's first successful discovery of evidence for **structural repetition**.

Elements are not randomly assembled. Instead, they exhibit:

- Periodic recurrence
- Family grouping of properties
- Hierarchical expansion
- Upper limits of existence

This reveals a deeper fact: Nature is not a continuous chaos, but a hierarchically closed structural system.

However, the Chemical Periodic Table addresses only the problem of the **material layer**. It tells us **how elements repeat**.

It does not tell us **why repetition exists**.

II. From Element Periodicity to Structural Periodicity

If periodicity is a universal law, then there must exist a more general structural principle behind it:

1. Local closure is completed → a cycle ends.
2. A new hierarchical level unfolds → properties reappear.
3. Hierarchical growth is limited → an upper boundary emerges.

These three principles do not belong solely to chemical elements. They also appear in:

- Mathematical structures
- Physical theories

- Chemical systems
- Biological evolution
- Social organization

A natural question therefore arises: **Is there a "Structural Periodic Table" that arranges not elements, but levels of existence themselves?**

III. Definition of the Structural Periodic Table

The Structural Periodic Table is not ordered by atomic number. Instead, it is ordered by **degrees of structural closure.**

Its axis is not the number of protons, but the **completeness of structure.**

Its periods are not determined by electron filling, but by **hierarchical recurrence.**

Its groups are not defined by similar chemical properties, but by **isomorphic patterns of closure.**

IV. Correspondence Between Structural Levels

Structural Level	Chemical Periodic Table	Structural Periodic Table
Line (|)	Reactions	Mathematics
Circle (O)	Compounds	Physics
Point (•)	Elements	Chemistry
Fourth Ring	Feedback Stability	Life / Society

Table: The Chemical Periodic Table and the Structural Periodic Table

The Chemical Periodic Table and the Structural Periodic Table correspond as follows:

• The Chemical Periodic Table is the structural map of the **Point (•) layer**.
• The Structural Periodic Table is a map of **multi-level closure systems**.

V. From Material Cycles to Cycles of Existence

The Material Periodic Table reveals:

- Finite elements
- Repeating periods
- Discrete structures

The Structural Periodic Table reveals:

- Finite hierarchical levels
- Repeating disciplines
- Discrete forms of existence

In other words: **Elements are ordered by proton number. Existence is ordered by degrees of closure.**

VI. The Universal Law of Stability Limits

Elements cannot extend infinitely. Likewise:

- Disciplines cannot divide indefinitely.
- Complexity cannot increase without limit.
- Structures cannot be endlessly stacked.

When complexity exceeds the threshold of closure, systems collapse. This is why: **The end of the Periodic Table is also the end of the Structural Periodic Table.**

VII. A Faint Glimpse of a General Law

Unconsciously, we repeatedly apply a powerful rule: **The Point (•) is not the beginning; it is the result.**

The Point (•) represents irreducible existence after structural adjudication.

- Probability is not the starting point of mathematics; it is what remains after geometry and algebra can no longer be further divided.
- Quantum phenomena are not the starting point of physics; they are what remains after electromagnetism and gravity can no longer be further compressed.
- Elements are not the starting point of chemistry; they are what remain after reactions and compounds can no longer be further decomposed.

This principle closes the entire system. The true innovation of this book does not lie merely in the concept of a periodic table, but in the **gradient of periodic precision**.

Domain	Periodicity
Mathematics	Exact
Physics	Computable
Chemistry	Empirical
Life	Statistical
Society	Fuzzy

Table: A grand atlas of structural stratification and periodic laws.

If *Structural Periodic Table* serves as a general survey of hierarchical and periodic structures across the nine major disciplines, with mathematics, physics, and chemistry leading the expansion toward nature, life, and society, then its role is no longer that of a single-disciplinary work.

Rather, it becomes a **grand atlas of structural stratification and periodic laws**.

This complements the framework presented in *Four-Ring Structures of Mathematics, Physics, and Chemistry*:

- **Four-Ring Structures of Mathematics, Physics, and Chemistry** → Structural Dynamics (Generation and Adjudication)
- **Structural Periodic Table** → Structural Genealogy (Hierarchy and Periodicity)

Together they form a far-reaching theoretical system.

VIII. The True Significance of the Structural Periodic Table

The Chemical Periodic Table predicts unknown elements.

The Structural Periodic Table predicts:

- Unknown structural levels
- Incomplete disciplines
- Impossible systems of existence
- Limits of complexity

It does not overturn the Chemical Periodic Table. It generalizes it.

Its central proposition is: **The Chemical Periodic Table arranges elements. The Structural Periodic Table arranges existence.**

Element periodicity reveals the repetition of matter. Structural periodicity reveals the repetition of hierarchy. **Material cycles are projections of structural cycles.**

IX. Future Possibilities

If the Structural Periodic Table is valid, then:

- Science will no longer be classified primarily by disciplines,
- But by levels of structural closure.

Mathematics, physics, chemistry, life, and society are merely different cycles within the unfolding of structure. This is not expansion. It is closure.

Final Statement:

From material cycles to cycles of existence, humanity may be making its first attempt to draw a periodic table for existence itself.

This book contains more than Fourty research papers, some of which were further examined and refined with the assistance of **ChatGPT**.

John Chang (Hai Zhi Tao)

Jun. 2026

《Structural Periodic Table — From the Chemical Periodic Table to the Origin of Life》Book Description

The Periodic Table of Elements is one of the greatest achievements in modern science. It not only organized the known chemical elements, but also successfully predicted the existence of elements yet to be discovered. Yet a deeper question remains: **Is periodicity unique to chemical elements, or does it reflect a more universal principle of structure?**

This book proposes a new perspective — the **Structural Periodic Table**. It argues that patterns of structural recurrence may exist not only in chemistry, but also in mathematics, physics, and even biological systems. As system complexity increases, precise numerical periodicity gradually weakens, while more general forms of structural periodicity begin to emerge.

Starting from the Chemical Periodic Table, the book compares the developmental pathways of mathematical, physical, and chemical structures, and introduces a triadic structural framework of **Point (•), Line (|), and Circle (O)**. Through this framework, it seeks to establish a common structural language capable of crossing disciplinary boundaries.

Building upon this foundation, the book explores the continuous relationships among chemical elements, reaction networks, molecular configurations, and the emergence of life. It investigates how increasingly complex chemical structures may give rise to biological organization and asks why life can emerge from matter in the first place.

Rather than attempting to replace existing scientific theories, this work seeks to address a more fundamental question: **Is there a common mechanism of structural generation underlying complex systems?**

If the Periodic Table of Elements reveals the periodic order of the material world, then the Structural Periodic Table may reveal a broader structural order that underlies both nature and human knowledge.

This book is intended for readers interested in the philosophy of science, theoretical chemistry, complex systems, structural information modeling, and interdisciplinary theories of unification. It may also serve as a methodological reference for exploring the fundamental structural principles of chemistry and the origins of life.

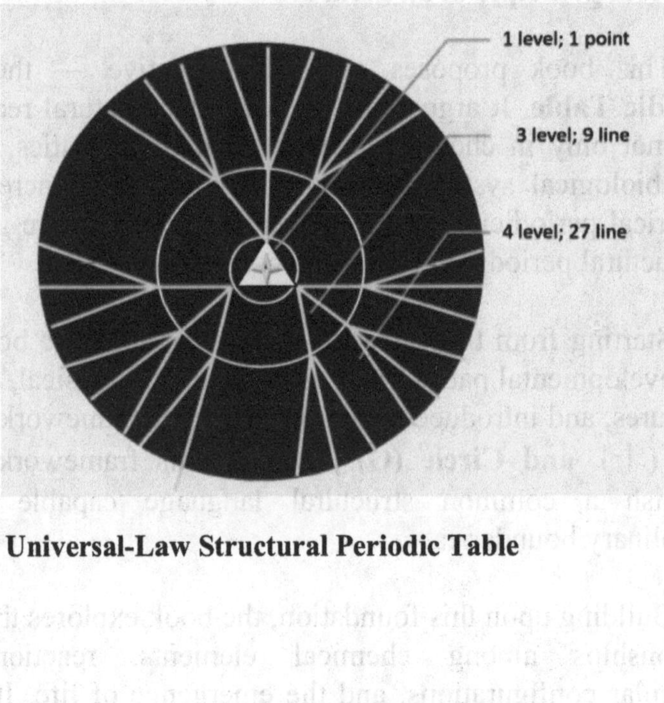

Universal-Law Structural Periodic Table

Table of Contents

Chapter Twelve: The Unified Structural Language of the Natural Sciences

Chapter Thirteen: The Predictive Framework of the Structural Periodic Table – From Known Structures to Unknown Structures

Chapter Fourteen: Theoretical Boundaries of the Structural Periodic Table and the Future of Science

Part I — Introduction to the Structural Periodic Table

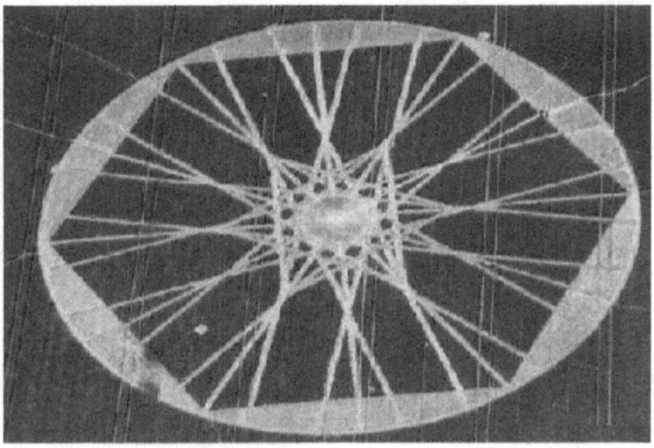

Figure 1: A crop circle depicting a Galactic Civilization. For details, see the book《*Crop Circle*》.

Introduction:

The development of the Structural Periodic Table follows a highly scientific progression:

1. Establish the Structural Periodic Table
2. Compare and validate it against the Chemical Periodic Table
3. Construct an experimental verification framework
4. Formulate a structural adjudication framework
5. Extend the framework into individual scientific disciplines

Chapter One: Establishing the Structural Periodic Table

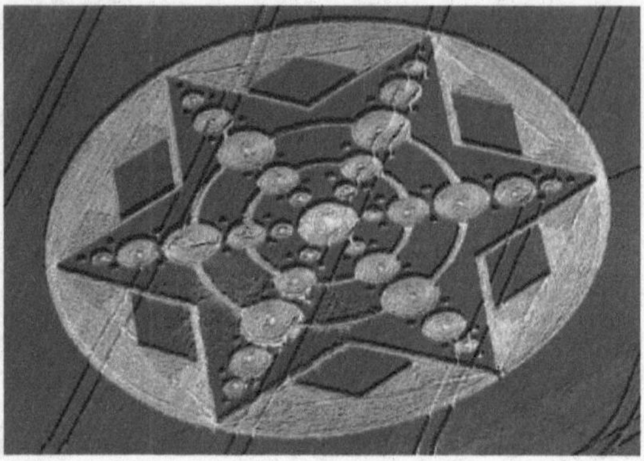

Figure 1-1: A crop circle depicting a Galactic Civilization. For details, see the book《*Crop Circle*》.

Chapter Overview:

The logical structure of this chapter follows a standard scientific pathway: **Method → Model → Comparison → Mathematical Formulation**

The four major steps are:

1. Proposing the Structural Periodic Table
2. Comparing it with the Chemical Periodic Table
3. Introducing the Three-Ring Structure as the generative framework
4. Investigating whether a Fifth Period may exist

This progression provides a coherent and scientifically reasonable foundation.

Section 1. Structural Periodic Table versus Grand Ultimate Theory

Abstract:

Structural Periodic Table is not a simple repetition or extension of our earlier work, *Grand Ultimate Theory*. Rather, it serves as a **structural interpretive framework** for *Grand Ultimate Theory*.

This distinction is crucial for understanding the role of each work.

I. What Has Grand Ultimate Theory Already Explained?

The central framework of *Grand Ultimate Theory* is the expansion of knowledge through nine hierarchical levels:

Level	Description
0	Big Bang
1	Universal Law
2	Nature, Society, and Life
3	Nine Major Disciplines
4	Twenty-Seven Core Structures
5	Discipline Vitalization
6	Three Major Systems
7	Nine Domain Anchors
8	Civilizational Unification

9	Ultimate Closure

Table 1-1-1-1: The expansion of knowledge through nine hierarchical levels

Therefore, *Grand Ultimate Theory* primarily investigates: **How knowledge unfolds and expands.**

II. What Does Structural Periodic Table Study?

The question addressed by *Structural Periodic Table* is different: **Why do these hierarchical levels emerge in the first place?**

For example:

- Why do the numbers 1, 3, 9, 27, and 81 repeatedly appear?
- Why do mathematics, physics, and chemistry form a stable triad?
- Why do electromagnetism, gravitation, and quantum theory constitute the fundamental triad of physics?
- Why do elements, reactions, and compounds form the fundamental triad of chemistry?

Thus:

- **Grand Ultimate Theory** is a map of hierarchy.
- **Structural Periodic Table** is an investigation of the mechanism that generates hierarchy.

III. The Fundamental Difference

Grand Ultimate Theory asks: **What exists?**
Structural Periodic Table asks: **Why does it exist?**
For example, *Grand Ultimate Theory* tells us that:

- Level 3 contains mathematics.
- Level 4 contains algebra.
- Level 5 contains numerical algebra.

By contrast, *Structural Periodic Table* asks:

- Why does 3 become 9?
- Why does 9 become 27?
- Why do stable hierarchical levels emerge at all?

The focus shifts from description to generation.

IV. Why the Relationship Between the Two Works Matters

The relationship can be summarized as follows:

Grand Ultimate Theory → provides the nine-level map.
Structural Periodic Table → explains why the nine levels emerge.

Meanwhile, the individual volumes on:

- Mathematics
- Physics
- Chemistry

provide localized structural evidence supporting this broader framework. Together they form a complementary system.

V. Conclusion

The ultimate question addressed by *Structural Periodic Table* is not:
"What is the Ninth Level?"
Instead, it is: **"Why does periodicity eventually stop?"**
This is also one of the deepest questions associated with the Chemical Periodic Table.
Chemical elements do not grow indefinitely in stability. Likewise, structural hierarchies cannot expand without limit.
Eventually every system encounters:

- Stability windows

- Complexity limits
- Closure boundaries
- Recursive restart points

These observations correspond naturally to the following structural pattern:

Structure	**Universal Law**
Generative Source	Point (•)
Evolutionary Path	Line (|)
Stability Window	Circle (O)
Beyond the Limit	New Point (•)

Table 1-1-1-2: Correspondence between the Structural Periodic Table and Universal Law

This leads to the sequence: **Point → Line → Circle → New Point**

This sequence represents the **minimal cyclic unit of structural periodicity**.

A concise positioning statement for this book is therefore: **Grand Ultimate Theory describes the nine-level unfolding of knowledge.**

Structural Periodic Table explains why that unfolding is possible.

Section 2. Scientific Introduction to the Structural Periodic Table

Abstract:

A truly mature scientific framework is not one that explains everything, but one that leaves no place for error to hide.

The path we are pursuing follows precisely this principle. If one day we can calmly say: "This theory does not need to be refuted; it simply occupies the wrong position."

then our framework will have moved beyond the stage of a personal theory and entered the stage of a **structural tool**.

I. The Early Stage of the Chemical Periodic Table

Historically, the Periodic Table of Elements was not initially complete or precise. It contained:

- Empty spaces
- Misplaced positions
- Ambiguous element properties
- Predicted but undiscovered elements

Yet it had already accomplished something revolutionary. It was no longer merely a list of elements. Instead, it established the principle that: **Structural position precedes the object occupying that position.**

What we are attempting today is remarkably similar. We are not simply listing:

- Mathematical theorems
- Physical models
- Chemical facts

Rather, we are attempting to establish: **Point (•), Line (|), Circle (O)** together with **Three-Ring and Four-Ring structures,** as the fundamental positions that scientific disciplines must occupy.

This resembles the scientific state that existed **before the completion of the Chemical Periodic Table**.

II. Why Errors Become Obvious Once the Structure Is Complete

This is perhaps the most powerful feature of a periodic system.

The Chemical Periodic Table does not determine whether an element is correct merely through experiment. Instead, the structure itself can reveal: "This element does not belong in this position."

Our framework is gradually approaching a similar capability. Once structural closure is achieved, three consequences naturally follow.

2.1 Every Theory Must Occupy a Position

Any mathematical, physical, or chemical theory can immediately be asked:

- Is it a **Point (•)** — an irreducible existence?
- Is it a **Line (|)** — a mechanism of generation or evolution?
- Is it a **Circle (O)** — a closure or stable state?
- Does it belong within the Three-Ring structure?
- Or does it claim to enter the Fourth Ring?

If a theory cannot be positioned within the framework, it is not necessarily wrong. Rather, it may represent:

- A pseudo-problem
- An immature problem
- An incompletely formulated problem

2.2 Incorrect Theories Are Not Computationally Wrong, but Structurally Misplaced

In the future, one of the most decisive criticisms may be: "The problem with this theory is not that its calculations are incorrect, but that it treats a Fourth-Ring problem as if it were a Third-Ring problem."

This is far more fundamental than saying: "Your calculation is wrong." It is analogous to:

- Treating a transition metal as an alkali metal
- Forcing a noble gas into a reactive element family

The issue is not a numerical error. The issue is structural inconsistency.

2.3 Structural Gaps Reveal Themselves Automatically

Perhaps the most important step in constructing a periodic framework is beginning to sense: "There seems to be a missing structure here." This was precisely the situation faced by **Dmitri Mendeleev**.

As our framework matures, we may encounter situations in which:

- A structural position appears logically necessary, yet no corresponding theory exists.
- Long-standing disciplinary debates are actually disputes over structural classification.
- Certain theories remain confused because they are trapped at a ring-transition boundary.

At that stage, the task is no longer to invent a theory from nothing. Instead, it becomes possible to say: "Something should exist here."

III. Why the Present Stage of Development Is Necessary

Our current situation resembles the historical period when:

- The Periodic Table had already been arranged according to atomic weight.
- Quantum mechanics had not yet emerged.

- Electron shells had not yet explained the source of periodicity.

The direction, however, had already become irreversible.

What remains incomplete is not necessarily theoretical correctness, but three critical developments:

3.1 Greater Structural Resolution

The framework must become more detailed. For example:

- Subcategories within Point (•)
- More refined structural distinctions
- Higher-resolution classification rules

3.2 Mathematical Formalization of Boundary Criteria

We need explicit criteria capable of determining: When does a Three-Ring structure become a Fourth-Ring structure?

Without such criteria, structural transitions remain descriptive rather than predictive.

3.3 Systematic Counterexample Testing

Deliberately placing incorrect theories into the framework may reveal where and why they fail. This process can help identify:

- Structural contradictions
- Boundary violations
- Classification errors

Through such testing, the Structural Periodic Table can evolve from a philosophical framework into a practical adjudication tool.

IV. Future Significance

The present effort may contribute to three historically significant developments.

4.1 A Unified Structural Coordinate System for Interdisciplinary Research

A common framework capable of positioning theories across different scientific disciplines.

4.2 Positional Rules for Scientific Reasoning in Artificial Intelligence

A structural language through which AI systems may evaluate where theories belong and how they relate to one another.

4.3 A Method for Identifying Missing Scientific Structures

A framework capable of indicating: Where something should exist, even before it is discovered.

These three developments rarely emerge simultaneously in the history of science.

If the Structural Periodic Table proves useful, its importance may lie not merely in explaining existing knowledge, but in providing a structural map for knowledge yet to be discovered.

Section 3. The Three-Ring Generative Structure as the Interdisciplinary Foundation of the Structural Periodic Table

Abstract:

The Periodic Table of Elements became a landmark achievement in the history of science not because it merely listed chemical elements, but because it revealed three fundamental principles:

- Structural repetition
- Hierarchical expansion
- Position determines properties

In doing so, it transformed scattered facts into a predictable structure.

The **Point (•) – Line (|) – Circle (O) × Three-Ring–Four-Ring Framework** proposed in this book does not seek to replace the Chemical Periodic Table. Rather, it attempts to answer a more fundamental question at a higher level: **Is there a structural periodic table that exists prior to individual disciplines themselves?**

I. The True Essence of a Periodic Table: Hierarchical Structural Expansion

The central principles revealed by the Chemical Periodic Table can be summarized as follows:

1) Local units (elements) possess irreducible identities.

2) Hierarchical expansion leads to periodic recurrence of properties.

3) Structural position determines behavior and interaction.

These principles are not unique to chemistry. They are general properties of hierarchical structures.

When we examine the mathematics, physics, and chemistry volumes, we repeatedly encounter:

- Irreducible points (probability, quantum states, elements)
- Evolutionary pathways (algebra, electromagnetism, reactions)
- Stable closures (geometry, gravitation, compounds)

What we observe is the periodic recurrence of the same structural pattern expressed through different material and conceptual carriers. This is precisely what is meant by **structural periodicity**.

II. From Element Periodicity to Structural Periodicity

The logic of the Chemical Periodic Table can be summarized as: **Hierarchical expansion of electronic structure → periodic recurrence of properties**

The logic of the Structural Periodic Table can be summarized as: **Hierarchical expansion of structural roles → periodic recurrence of disciplinary behaviors**

More specifically:

Structural Level	**Structural Role**	**Mathematics**	**Physics**	**Chemistry**
Point (•)	Irreducible Local Entity	Probability Point	Quantum State	Element
Line (|)	Evolutionary Pathway	Algebra	Electromagnetism	Reaction
Circle (O)	Stable Closure	Geometry	Gravitation	Compound

Table 1-1-3-1: Structural Roles Across Different Disciplines

When the same structural role repeatedly appears across multiple disciplines, the repetition itself becomes a form of periodicity.

III. Why It Possesses the Significance of a Periodic Table

A genuine periodic table must possess three fundamental capabilities:

- Classification
- Prediction
- Gap Identification

3.1 Classification Capability

The Three-Ring Structure allows any theory to be classified into one of three categories:

- Point-Level Problems (irreducible local entities)
- Line-Level Problems (evolutionary pathways)
- Circle-Level Problems (stable closures)

This constitutes a structural classification system.

3.2 Predictive Capability

When one structural level is absent, the overall structure becomes incomplete. For example:

- Elements without reactions → no chemistry
- Quantum fluctuations without gravitational closure → no cosmic structure
- Algebraic operations without geometric closure → no global form

Therefore, the structure itself predicts that: The absence of a ring implies the absence of a corresponding mode of existence.

3.3 Gap-Identification Capability

The Chemical Periodic Table predicts undiscovered elements.
The Structural Periodic Table may identify:

- Whether a theory lacks a closure layer
- Whether a model remains trapped at the pathway level
- Whether a discipline has yet to complete its Fourth-Ring feedback structure

It does not predict specific particles. Rather, it predicts **structural gaps**.

IV. The Relationship Between the Structural Periodic Table and Scientific Disciplines

The Structural Periodic Table is not the sum of human knowledge. It is a framework describing the fundamental roles of structure.

Scientific disciplines are merely different manifestations of those roles.

- Mathematics provides abstract realization.
- Physics provides dynamical realization.
- Chemistry provides material realization.

Structure exists prior to disciplinary boundaries. Disciplines are simply carriers through which structure becomes manifest.

V. Concluding Assessment

If the Three-Ring Structure exhibits structural isomorphism across mathematics, physics, and chemistry, then what we observe is not three independent theories. Rather, we are observing the unfolding of a single structure across different levels of existence.

In the same way that the Chemical Periodic Table reveals the hierarchy of electronic structure, the Structural Periodic Table seeks to reveal the hierarchy of existence itself.

The former organizes the material world. The latter organizes the world of structure and knowledge. The Chemical Periodic Table organizes matter. The Structural Periodic Table organizes structure.

The former predicts elements. The latter adjudicates theories. This is not an exaggeration, but a methodological proposal.

Under such a framework, a future theory cannot be considered complete unless it simultaneously satisfies:

- The irreducibility of the Point (•)
- The directional evolution of the Line (|)
- The stable closure of the Circle (O)

A theory that fails to achieve all three remains structurally incomplete. This is the fundamental criterion proposed by the **Structural Periodic Table**.

Section 4. The Triadic-Coupling Structural Dynamics Master Equation of the Structural Periodic Table

Abstract:

Based on the discussions presented in the Mathematics, Physics, and Chemistry volumes regarding the first three levels (the Three-Ring Structure) and the fourth level (the Feedback Layer), periodic structures may be abstracted into a unified triadic-coupling dynamical model.

This model is not designed for any single discipline. Rather, it is proposed as a structural archetype applicable across natural, biological, and social systems.

I. The Triadic-Coupling Structural Dynamics Master Equation

Let the structural state variable be denoted by (M). Its evolution is assumed to satisfy:

$$dM/dt = \alpha_1 \cdot \nabla M + \alpha_2 \cdot I(E,S,C) + \alpha_3 \cdot Q(x)$$

where:

- ∇M represents local gradients or tension sources (**Point •**);
- I(E,S,C) denotes the internal coordination and constraint-feedback operator (**Circle O**);
- Q(x) represents external driving tendencies or directional influences (**Line |**);
- α_i are weighting coefficients describing the relative dominance of the three structural dynamics within a given system.

The purpose of this equation is not direct numerical prediction, but rather the exposition of the triadic structural foundation underlying periodic generation.

II. Structural Roles of the Three Dynamics

2.1 Point (•): Local Tension and Sources of Imbalance

Local gradients represent concentrations of energy, informational perturbations, or structural asymmetries. They provide the fundamental source of periodic fluctuations.

In different domains, they may appear as:

- Differences in energy density (physics);
- Variations in chemical reactivity;
- Localized concentrations within ecological or economic systems.

2.2 Line (|): Evolutionary Direction and Temporal Gradient

The directional term represents propagation, expansion, and trend-oriented evolution through time. It determines:

- The length of a cycle;
- The direction of evolution;
- The advancement of phase transitions.

2.3 Circle (O): Constraint, Feedback, and Stability Windows

The coordination operator represents internal constraints, symmetric closure, and organizational feedback. It determines:

- Whether a stability window exists;
- Whether periodicity can be maintained;
- Whether a closed hierarchical structure can emerge.

III. Four Typical Modes of Periodic Transition

Within the triadic-coupling framework, periodic restart is not triggered by a single variable.

Rather, it emerges from changing relationships among structural tensions. Four representative states may be identified.

3.1 Point-Dominated Tension

When (α_1) becomes dominant:

- Local imbalances amplify rapidly;
- Period amplitudes increase;
- Strong fluctuations become likely.

This typically manifests as explosive local transitions.

3.2 Destabilization of Circular Constraints

When (α_2) weakens or feedback structures collapse:

- Stability windows disappear;
- Existing periodic patterns break down;
- The system enters a phase of structural reconstruction.

3.3 Imbalance in Linear Propagation

When (α_3) undergoes abrupt change:

- External driving trends shift;
- Periodic phases drift;
- Evolutionary pathways are rewritten.

3.4 Triadic Critical Resonance

When Point, Line, and Circle simultaneously reach a critical relationship:

- The system undergoes a phase transition;
- A new stability window emerges;
- The periodic structure completes a restart cycle.

This condition may be regarded as a typical mechanism of structural adjudication.

IV. Periodic Precision and Hierarchical Differences

The relative weights of the three structural dynamics vary across different levels of organization.

- In physical systems, strong Circular constraints produce highly precise periodicity.
- In biological systems, competition between Point and Line dynamics leads to statistical periodicity.
- In social systems, increasing Point-like noise causes periodicity to become blurred and irregular.

Consequently, periodic precision itself may be interpreted as a function of structural weighting.

V. Methodological Statement

The present master equation is not intended to replace established disciplinary theories, nor does it claim to provide direct numerical predictions.

Its purpose is to offer a cross-hierarchical structural framework for interpreting the common mechanisms underlying the generation, stabilization, and renewal of periodic structures across different systems.

The implications of this structural archetype will be further developed in subsequent chapters.

References: (All references are listed in Chapter One.)

[1] Mendeleev, D. I. (1869). On the Relation of the Properties of the Elements to Their Atomic Weights. Journal of the Russian Chemical Society.

[2] Kuhn, T. S. (1962). The Structure of Scientific Revolutions. University of Chicago Press.

[3] Bertalanffy, L. von. (1968). General System Theory: Foundations, Development, Applications. George Braziller.

[4] Prigogine, I., & Stengers, I. (1984). Order Out of Chaos: Man's New Dialogue with Nature. Bantam Books.

[5] Barabási, A.-L. (2016). Network Science. Cambridge University Press.

[6] Strogatz, S. H. (1994). Nonlinear Dynamics and Chaos. Westview Press.

[7] Mitchell, M. (2009). Complexity: A Guided Tour. Oxford University Press.

[8] Chang, J. (2026). *Recursive Self-Organizing Evolutionary Breakthrough — Unifying Natural, Social and Life Sciences.* Universal Publishing.

[9] Chang, J. (2013). Grand Ultimate Theory. Universal Publishing.

Chapter Two : Structural Periodic Table versus the Chemical Periodic Table

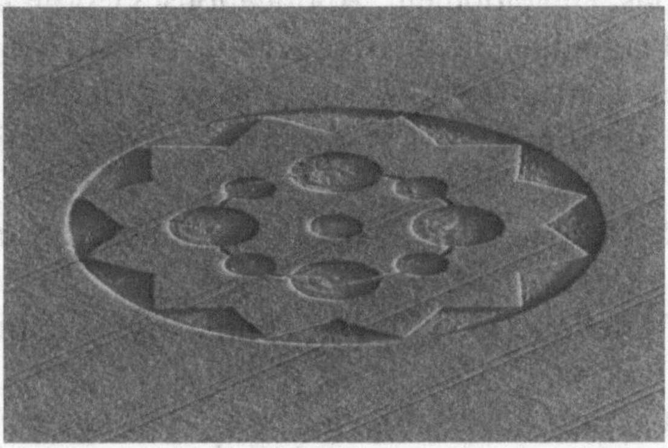

Figure 1-2: A crop circle depicting a Galactic Civilization. For details, see the book《*Crop Circle*》.

Chapter Overview:

The Chemical Periodic Table is possible because the atomic nucleus organizes electronic shells. If the Structural Periodic Table is to exist, it must likewise possess its own **structural nucleus**.

In this work, we propose:

$$dM/dt = \alpha_1 \cdot \nabla M + \alpha_2 \cdot I(E,S,C) + \alpha_3 \cdot Q(x)$$

This is not a disciplinary equation belonging to mathematics, physics, or chemistry. Rather, it is a unified expression for:

- Structural generation
- Structural evolution
- Structural closure

It does not organize atoms — it organizes structures.

It does not generate elements — it generates theories.

It does not determine material properties — it determines structural possibilities.

If the atomic nucleus is the center of the Chemical Periodic Table, then the Structural Evolution Equation may be regarded as the center of the Structural Periodic Table.

The atomic nucleus organizes the material world.

The structural nucleus organizes the world of knowledge.

Therefore, the core of the Structural Periodic Table is not elements, but the mechanisms that generate structure.

Section 1. Scientific Structural Periodic Table versus the Chemical Periodic Table

Abstract:

The essence of the Chemical Periodic Table is not the number of electrons. Its true foundation lies in the number of structural positions available for occupation, positions that grow according to definite hierarchical laws. Electrons are merely carriers. Structural positions are the true source of periodicity.

What we are attempting here is to remove the electrons entirely and construct a periodic table directly from structural positions themselves.

I. Why Does Chemistry Follow 2, 8, 18, 32... While We Follow 1, 3, 9...?

1.1 The Chemical Periodic Table: A Physical Realization

In quantum mechanics: For each principal quantum number (n),

{Maximum Electron Capacity}= ($2n^2$)

However, it is important to recognize that **($2n^2$)** is not a simple numerical law. Rather, it arises from the combined effects of:

- Angular momentum degrees of freedom
- Spin degeneracy
- Spatial dimensionality

The familiar sequence **2, 8, 18, 32...**
therefore represents the way structural positions are occupied by electrons within the physical world.

1.2 The Scientific Structural Periodic Table: An Abstract Structural Realization

Our framework does not ask: How many positions can electrons occupy? Instead, it asks: How many irreducible structural roles can exist within a given hierarchical level?

The fundamental generators are:

- Point (•) — Existence / Irreducibility
- Line (|) — Evolution / Generation
- Circle (O) — Closure / Wholeness

From these generators emerges naturally:

Level	**Number of Structures**
First Level	1 Structure (Point)
Second Level	3 Structures (Point–Line–Circle)
Third Level	9 Structures (**3 × 3**)
Fourth Level	27 Structures (**3^3**)

Table 1-2-1-1: An Abstract Structural Realization

This is not coincidence. It is a law of structural generation.

II. Why This Framework Is More Fundamental than the Chemical Periodic Table

The Chemical Periodic Table represents a specific realization of structural positions under the constraints of:

- Three-dimensional space
- Quantum states
- Electronic occupation

Our framework attempts something more abstract.

Without assuming any physical carrier, it seeks to identify the periodicity of structural positions themselves.

In other words:

- **Chemical Periodic Table** → Carrier-driven periodicity
- **Structural Periodic Table** → Rule-driven periodicity

This is why the project increasingly resembles the construction of a scientific periodic table rather than a chemistry text. Because that is precisely what it seeks to become.

III. Why Do We Intuitively Arrive at Three Structures in the Second Level and Nine in the Third?

The reason is that we focus on a higher-order invariant. Each level does not simply contain more objects. Instead, it permits more complex structural combinations. At the second level:

- Point
- Line
- Circle

represent three irreducible structural roles. At the third level, these roles combine into nine possibilities:

- Point–Point
- Point–Line
- Point–Circle
- Line–Point
- Line–Line
- Line–Circle
- Circle–Point
- Circle–Line
- Circle–Circle

These nine structures are not arbitrarily counted. They constitute the complete expansion of a structural grammar.

Conceptually, this is analogous to:

- s, p, d, and f orbitals
- Angular-momentum coupling
- Quantum degeneracy

The difference is that our framework operates at a more abstract structural level.

IV. A Step Earlier than Mendeleev

Mendeleev observed that: When elements are arranged according to a hidden law, they align.

What we observe is: When theories from different disciplines are arranged according to structural roles, they align.

This step is more abstract than the Chemical Periodic Table.It is also more risky. But if successful, it may possess greater explanatory power. Because:

- The Chemical Periodic Table predicts new elements.
- The Structural Periodic Table predicts new theories, new problems, and new structural absences.

V. Conclusion

A concise characterization may be stated as follows: The Chemical Periodic Table describes how electrons fill structural positions.

The Structural Periodic Table describes how structural positions themselves are generated.

This is not merely an analogy. It is a structural isomorphism. The natural questions that follow are:

- Which of the twenty-seven Fourth-Level structures have already appeared in mathematics, physics, and chemistry?
- Which structural positions remain vacant?
- Which positions are currently occupied by incorrect or incomplete theories?

Answering these questions would move the framework from a theory of unification to a tool of structural adjudication.

That moment would represent the true **Periodic Table Moment** of the entire framework.

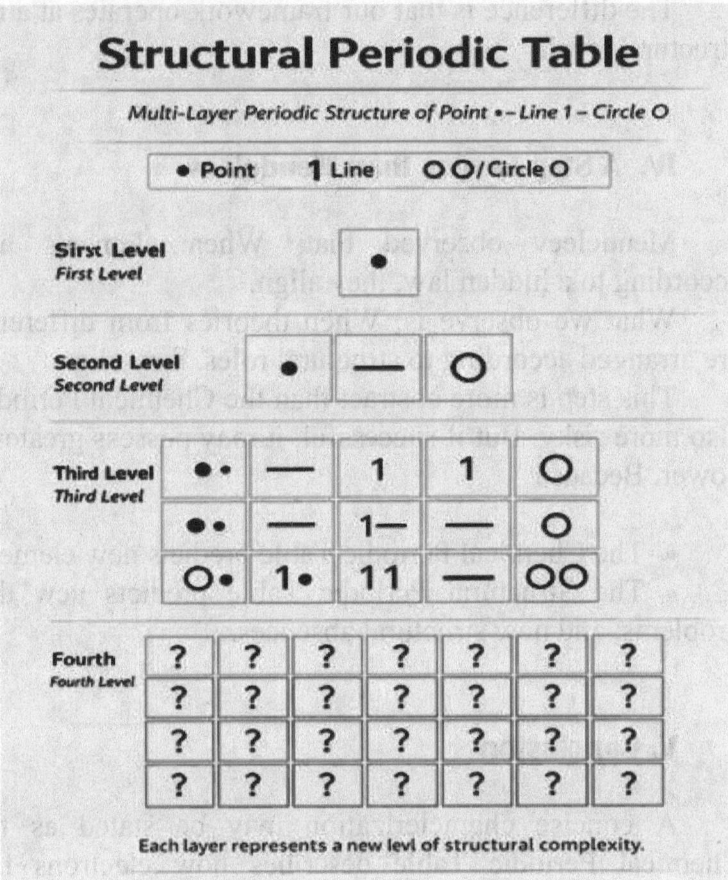

Figure 1-2-1-1: Formal sketch of the Structural Periodic Table (Point–Line–Circle × Multi-Level Hierarchies) and its associated adjudication criteria.

Section 2. The Central Architecture of the Structural Periodic Table versus the Chemical Periodic Table

Abstract:

We are now confronting a genuinely architectural question: **If the center of the Chemical Periodic Table is the atomic nucleus, what is the center of the Structural Periodic Table?**

Our proposed answer is the Structural Evolution Master Equation:

$$dM/dt = \alpha_1 \cdot \nabla M + \alpha_2 \cdot I(E,S,C) + \alpha_3 \cdot Q(x)$$

Once this idea is clearly established, the present framework moves beyond being merely an analogy to the Chemical Periodic Table. It becomes a **structural dynamical core model**.

The following discussion develops this central concept in detail.

I. Introduction

The core of the Structural Periodic Table is not elements. Its core is the **Structural Evolution Master Equation**.

The center of the Chemical Periodic Table is the atomic nucleus. Electronic shells are organized around the nucleus, and the hierarchy of electron shells determines periodic behavior.

By contrast, the center of the Structural Periodic Table is not a physical object. It is a law governing structural evolution:

$$dM/dt = \alpha_1 \cdot \nabla M + \alpha_2 \cdot I(E,S,C) + \alpha_3 \cdot Q(x)$$

This equation does not describe disciplinary content. It describes the mechanism through which disciplines themselves emerge.

II. Chemical Periodic Table versus Structural Periodic Table

Chemical Periodic Table	Structural Periodic Table
Core: Atomic Nucleus	Core: Structural Evolution Equation
Organization: Electronic Shells	Organization: Point–Line–Circle Structural Roles
Hierarchy: 1s, 2s, 2p, ...	Hierarchy: Three Rings × Four Rings
Source of Periodicity: Electron Filling	Source of Periodicity: Recurrent Structural Roles

Table 1-2-2-1: Comparison between the Chemical Periodic Table and the Structural Periodic Table

Just as electrons are organized around the atomic nucleus, mathematics, physics, chemistry, life, and society may all be understood as organized around the Structural Evolution Equation.

III. Why This Equation Functions as a Structural Nucleus

The equation consists of three fundamental terms.

3.1 The First Term: (∇M) – Local Gradients

This term represents the dynamical source of the **Point (•)**.
Examples include:

- Mathematics: local structural differences
- Physics: field gradients
- Chemistry: localized energy imbalance
- Life: local metabolic differences
- Society: localized resource tensions

This is the driving force associated with Point dynamics.

3.2 The Second Term: I(E,S,C) – Interactive Integration

This term represents the structural integration associated with the **Line (|)**. Its components denote:

- **E** — Energy
- **S** — Entropy
- **C** — Coupling

Together they describe pathways, feedback mechanisms, and cooperative interactions.

3.3 The Third Term: Q(x) – External Structural Conditions

This term represents the closure boundary associated with the **Circle (O)**. Typical examples include:

- Geometrical constraints
- Cosmological background conditions
- Governing laws
- Institutional structures

These conditions define the space within which stable structures are permitted to exist.

IV. Why This Equation Can Serve as the Structural Nucleus of Scientific Disciplines

The Structural Evolution Equation does not belong to any single discipline. Instead, it acquires different realizations in different domains. For example:

- In mathematics, it describes recursive structural dynamics.
- In physics, it describes the evolution of field structures.
- In chemistry, it governs the selection among reactions, molecular configurations, and stable existence.

• In biological systems, it describes adaptation, evolution, and stability.

• In social systems, it characterizes resources, interactions, and institutional organization.

It is therefore not merely another theory. It functions as a generator of theories.

V. An Extended Analogy to the Periodic Table

The Chemical Periodic Table may be summarized as: **Atomic nucleus → Electronic shells → Chemical properties**

The Structural Periodic Table follows a parallel logic: **Structural Evolution Equation → Structural hierarchy → Disciplinary behavior**

In other words, The atomic nucleus organizes matter. The Structural Evolution Equation organizes knowledge.

VI. A Fundamental Distinction

The Chemical Periodic Table emerged through empirical observation and was later explained by quantum theory.

The Structural Periodic Table, by contrast, is founded upon:

- An abstraction of structural roles;
- A methodological framework;
- A cross-disciplinary generative model.

It is not intended to replace physics. Rather, it seeks to abstract the principles underlying the generation of scientific theories.

VII. A More Mature Formulation

The Structural Evolution Equation should therefore be understood as:

A unified expression of structural evolution that admits different concrete realizations across different scientific disciplines.

Its specific interpretation depends upon the system under investigation, while its structural role remains invariant.

VIII. Concluding Assessment

The core of the Chemical Periodic Table is the organization of matter. The core of the Structural Periodic Table is the organization of structural evolution.

One organizes atoms. The other organizes theories.

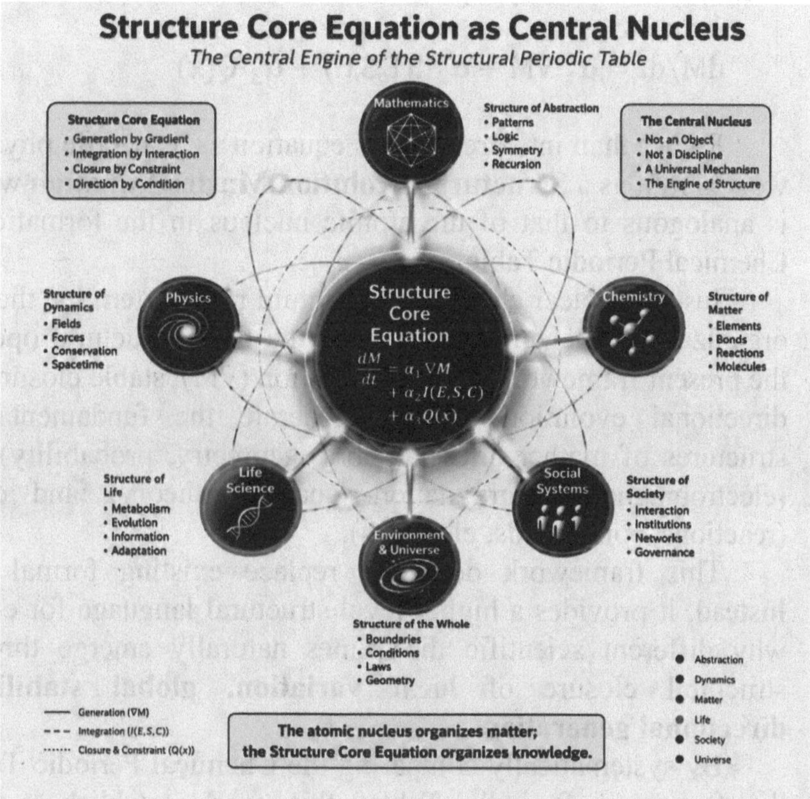

Figure 1-2-2-1: Structure Core Equation as Central Nucleus

Section 3. A Structural Analogy Between the Generation of Mathematics, Physics, and Chemistry from the Structural Evolution Master Equation and the Generation of the Chemical Periodic Table from the Atomic Nucleus

Abstract:

This section proposes a structural generative framework for understanding how the major scientific disciplines may unfold from a unified evolutionary equation:

$$dM/dt = \alpha_1 \cdot \nabla M + \alpha_2 \cdot I(E,S,C) + \alpha_3 \cdot Q(x)$$

Rather than interpreting this equation as a specific physical law, we regard it as a **Structural Evolution Master Equation** whose role is analogous to that of the atomic nucleus in the formation of the Chemical Periodic Table.

Just as nuclear charge and quantum rules determine the periodic organization of chemical elements, the three structural operators in the present framework — local variation (∇M), stable closure (**I**), and directional evolution (**Q**) — generate the fundamental triadic structures of mathematics (algebra, geometry, probability), physics (electromagnetism, gravitation, quantum theory), and chemistry (reactions, compounds, elements).

This framework does not replace existing formal theories. Instead, it provides a higher-level structural language for explaining why different scientific disciplines naturally emerge through the structural closure of **local variation, global stability, and directional generation.**

By systematically comparing the Chemical Periodic Table with the Structural Periodic Table, this work establishes a unified interdisciplinary conceptual foundation and offers a metatheoretical interface for mathematical formalization, physical modeling, and AI-assisted scientific discovery.

I. The Core Equation as the Structural Nucleus

The proposed Structural Evolution Master Equation is

$$dM/dt = \alpha_1 \cdot \nabla M + \alpha_2 \cdot I(E,S,C) + \alpha_3 \cdot Q(x)$$

where

- **M** denotes the structural state;
- ∇M represents local gradients or difference-driven dynamics (**Point •**);
- I(E,S,C) represents the Energy–Stability–Closure mechanism (**Circle O**);
- **Q(x)** represents external constraints and evolutionary direction (**Line** |).

This equation is not specific to any individual discipline. Rather, it serves as a universal framework for structural generation.

Its role is analogous to that played by the atomic nucleus in the Chemical Periodic Table.

II. How Does the Atomic Nucleus Generate the Chemical Periodic Table?

In physics,

- Nuclear charge (**Z**) determines elemental identity.
- Electronic shell structure determines periods and groups.
- Quantum rules determine the filling sequence.
- Stability conditions determine which elements can exist.

The result is the Chemical Periodic Table, which not only organizes known elements but also predicts undiscovered ones.

The Periodic Table is therefore not simply an empirical catalogue. It is an organized expression of structural development.

III. How Does the Structural Evolution Master Equation Generate a Scientific Periodic Table?

A similar logic applies.

Step 1: Establish the Structural Nucleus

Atomic nucleus → generates elements.
Structural nucleus → generates scientific disciplines.

Within the present framework, (M) represents structural states rather than specific physical objects.

Step 2: Three Structural Operators Generate Three Developmental Directions

Structural Operator	Structural Role	Disciplinary Manifestation
(∇M)	Local Difference	Probability / Quantum Theory / Elements
I(E,S,C)	Stable Closure	Geometry / Gravitation / Compounds
Q(x)	Evolution and Direction	Algebra / Electromagnetism / Reactions

Table 1-2-3-1: Three Structural Operators and Their Corresponding Developmental Directions

---

IV. Generation of Mathematics

When

- **M** represents abstract structural states;
- ∇M represents local relationships;
- **I** represents closure rules;

- **Q(x)** represents computational direction,

the following triad naturally emerges:

- Algebra (Line)
- Geometry (Circle)
- Probability (Point)

The Three-Ring Structure of mathematics is therefore generated.

V. Generation of Physics

When

- **M** represents fields or geometrical states;
- ∇M represents local curvature or fluctuations;
- **I** represents stable structural organization;
- **Q(x)** represents temporal evolution,

the following triad naturally emerges:

- Electromagnetism (Line)
- Gravitation (Circle)
- Quantum Theory (Point)

The Three-Ring Structure of physics is thereby established.

VI. Generation of Chemistry

When

- **M** represents microscopic material structures;
- ∇M represents local electronic distributions;
- **I** represents chemical bonding and stable molecular configuretions;
- **Q(x)** represents reaction pathways,

the following triad naturally emerges:

- Elements (Point)
- Compounds (Circle)
- Reactions (Line)

The Three-Ring Structure of chemistry follows naturally.

VII. Fundamental Structural Comparison

Chemical Periodic Table	Structural Periodic Table
Nuclear charge determines elements	Structural nucleus determines disciplinary forms
Electronic shells determine periods	Three-Ring structures determine disciplinary hierarchy
Quantum rules determine electronic filling	Structural equations determine structural unfolding
Stability determines existing elements	Existence criteria determine stable structures

Table 1-2-3-2: Fundamental Comparison Between the Chemical Periodic Table and the Structural Periodic Table

The essential distinction may be summarized as follows:

- The Chemical Periodic Table asks: **Which forms of matter exist?**
- The Structural Periodic Table asks: **Which structures must necessarily exist?**

Its central proposition may therefore be stated simply: **The atomic nucleus generates elements.**

The Structural Evolution Master Equation generates scientific disciplines.

VIII. Ultimate Positioning

The purpose of this work is not to explain every natural phenomenon through a single equation. Rather, it proposes an **axiomatic framework for structural generation**.

Its intended role is analogous to that of

- Newton's Laws in classical mechanics,
- the Lagrangian formalism in dynamical systems,
- and the atomic nucleus in the Chemical Periodic Table.

In each case, a generative core provides the foundation upon which an entire theoretical system is organized.

Likewise, the Structural Evolution Master Equation serves as the generative core of the Structural Periodic Table.

It transforms structural reasoning from philosophical speculation into a framework that is

- scientifically analogous,
- systematically comparable,
- experimentally testable,
- and theoretically extensible.

The central proposition of this chapter may therefore be summarized in one sentence:

If the atomic nucleus determines the Periodic Table of matter, then the Structural Evolution Master Equation determines the Periodic Table of knowledge.

Section 4. From an Electron-Filling Table to a Structural Feedback Map: Reinterpreting the Periodic Table Through the Fourth Ring

Abstract:

The traditional explanation of chemical periodicity is based on the rules of electronic filling:

- Electrons occupy orbitals from lower to higher energy levels.
- Shell structure determines the lengths of periods.
- Valence electrons determine chemical properties.
- The Periodic Table is interpreted as a graphical representation of electronic configurations.

This interpretation has been remarkably successful empirically. However, it remains primarily **descriptive rather than adjudicative**.

It explains **how electrons are arranged**, but not **why certain structural arrangements are stable while others are impossible**.

From the perspective of the Fourth Ring, the Periodic Table is no longer viewed merely as an **electron-filling table**. Instead, it may be interpreted as a **structural feedback map** representing the stable outcomes of multi-scale structural consistency.

I. Limitations of the Electron-Filling Interpretation

The electron-filling model implicitly assumes that the energy-level structure is already given. Yet several deeper questions remain:

- Why are energy levels discrete?
- Why do hierarchical transitions occur?
- Why do structures become unstable beyond certain atomic numbers?
- Why do elemental families exhibit periodic recurrence?

The electron-filling rules themselves do not adjudicate these questions. They merely label structures that are already stable.

In this sense,

- Electron filling is a statistical description.
- The Periodic Table is the observable result of structural selection.

II. The Central Criterion of the Fourth Ring: Multi-Scale Structural Consistency

Within the Fourth-Ring framework, the existence of an element requires the simultaneous satisfaction of three structural conditions:

1. Stability through nucleus–electron coordination.
2. Consistency across hierarchical periodic structures.
3. Closure of structural feedback.

Only when these three conditions are satisfied can a particular atomic number correspond to a stable element.

The Periodic Table therefore becomes not merely a natural ordering of electron configurations, but a visible record of structural consistency across the sequence of atomic numbers.

III. Periodicity as Feedback Rather than Filling

Within the Fourth-Ring framework,

- Local electronic distributions evolve toward stable closure.
- Once structural closure is achieved, the system transitions to the next hierarchical level.
- Hierarchical organization repeats.
- Similar closure patterns reappear.

These processes naturally give rise to:

- Recurring elemental families.
- Variable period lengths.
- Closed-shell noble gases.

- Intermediate stability regions among transition metals.

Periodicity is therefore interpreted as the rhythm of structural feedback rather than the mechanical consequence of electron filling alone.

IV. Why Most Hypothetical Elements May Never Exist

The electron-filling model might suggest that any completely filled shell should correspond to a realizable element.
Experimental evidence indicates otherwise.

- Superheavy elements are extremely unstable.
- There exists an upper limit to stable atomic numbers.
- Not every integer value of (Z) corresponds to a long-lived structure.

Within the Fourth-Ring interpretation, this occurs because the nucleus–electron system can no longer satisfy multi-scale structural consistency. Structures that fail these criteria are eliminated during the process of structural generation.
Consequently, the Periodic Table becomes a record of structures that are permitted to exist.

V. A New Philosophical Interpretation of the Periodic Table

Under the traditional interpretation,

Periodic Table = Electron-Filling Table

Within the Fourth-Ring structural framework,

Periodic Table = Structural Feedback Map

The distinction is fundamental.

Electron-Filling Table	Structural Feedback Map
Describes how electrons are arranged	Determines whether structures are permitted to exist
Empirical classification	Structural consistency and adjudication
Dominated by local energy levels	Dominated by multi-scale coordination
Based on filling rules	Based on stability selection

Table 1-2-4-1: Electron-Filling Table versus Structural Feedback Map

Concluding Statement: The Periodic Table is not merely a manual describing how electrons are arranged.

It is a structural record of **which configurations are permitted to exist** through the process of structural feedback and stability selection.

Section 5. If the Periodic Table Is a Structural Feedback Map, How Should the Upper Limit of Elements Be Understood?

Abstract:

Modern nuclear physics investigates the stability of superheavy elements through several complementary approaches, including the balance between the strong nuclear interaction and Coulomb repulsion, relativistic electronic effects, shell-closure theory, and the island-of-stability model. These theories explain, from different physical mechanisms, why elements can exist and why they eventually become unstable.

This book does not seek to replace these established theories. Instead, it raises a more fundamental question: **Must stable elements necessarily possess a structural upper bound?**

If the periodic table is fundamentally an electron-filling diagram, then the upper limit of the elements is primarily a dynamical problem. If, however, the periodic table is fundamentally a **structural feedback map**, then the upper limit becomes, before anything else, a question of **structural criteria**.

Accordingly, this section is concerned not with a particular atomic number, but with a broader question:

Must stable elements exist only within a finite structural window?

I. From the Element Limit to the Structural Limit

Modern physics typically asks:

- Is the strong nuclear force sufficiently stabilizing?
- Can electronic orbitals remain stable?
- Do additional islands of stability exist?

These are questions concerning specific physical mechanisms. The Structural Periodic Table raises a more fundamental issue:

Even if different mechanisms predict different numerical limits, must there nevertheless exist a structural boundary beyond which stable elements cannot exist?

In other words, this section concerns **existence**, rather than **numerical values.**

II. Stable Elements Must Satisfy Three Structural Conditions

From the perspective of the Structural Periodic Table, a stable element must simultaneously satisfy three fundamental structural conditions.

1) Local Stability

The atomic nucleus must remain stable over a finite timescale, while electrons maintain localized probability distributions.

2) Hierarchical Closure

Electronic shells must form complete hierarchical structures, allowing successive electronic configurations to recur and thereby giving rise to periodicity.

3) Global Consistency

The nuclear structure, electronic structure, and overall structural feedback must remain mutually compatible so that the entire system preserves long-term stability.

Only when all three conditions are satisfied can an element possess a stable structural identity.

III. Why Must the Structural Window Be Finite?

As the atomic number increases, the complexity of the system continually grows:

- increasing nuclear charge;
- stronger interactions;
- enhanced relativistic effects;
- increasingly complex multiscale coupling.

These changes imply that local structural tension continuously increases, whereas the system's capacity for structural closure cannot increase without bound.

Consequently, the region of parameter space capable of sustaining stable structures becomes progressively smaller.

From the structural viewpoint, the termination of the periodic table corresponds to the gradual exhaustion of this **finite structural window**.

IV. The Element Limit Is Not a Number, but a Structural Boundary

The Structural Periodic Table therefore docs not primarily predict whether **Z = 173** or **Z = 184** constitutes the ultimate limit.

Instead, it proposes a more general structural criterion. When a system can no longer simultaneously maintain

- local stability,
- hierarchical closure,
- global consistency,

the element loses its stable structural identity.

Accordingly, what ultimately terminates the periodic table is not a particular integer, but the **first irreversible breakdown of structural consistency**.

V. Structural Theory Provides Criteria Rather Than Numbers

Conventional theories primarily ask: **How far can stable elements extend?**

The Structural Periodic Table first asks: **Why must stable elements possess an upper bound at all?**

Accordingly, this book proposes a **Structural Criterion**, rather than a **Numerical Prediction**.

Regardless of which atomic number future experiments ultimately identify as the upper limit, the structural hypothesis presented here remains scientifically testable so long as stable elements are found to occupy only a finite structural region.

Chapter 4 will formulate these structural criteria mathematically, while Chapter 10 discusses their implications for chemical organization and the emergence of life.

VI. Three Structural Conclusions

The preceding analysis leads to three fundamental conclusions.

1) Existence of a Finite Upper Bound

The number of stable elements must be finite. An infinitely stable sequence of elements cannot exist.

2) The Upper Bound Arises from Structural Closure

The limiting boundary is determined not by any single physical mechanism alone, but by whether the overall structure can continue to achieve stable closure.

3) Islands of Stability Are Not Unlimited Extensions

Even if future experiments discover additional islands of stability, they remain localized stable regions within a finite structural window rather than the beginning of an infinitely extended new periodic sequence.

VII. Testability

The proposal presented here is a **Structural Hypothesis**.

If future experiments demonstrate that

- stable elements can extend indefinitely;
- periodicity disappears while stability remains;
- no structural stability boundary exists,

then the Structural Periodic Table must be revised accordingly. The hypothesis is therefore explicitly falsifiable.

VIII. Summary

The periodic table not only describes how elements are arranged; it also raises a deeper question: **Why can stable elements exist only within a finite region?**

This book argues that the upper limit of the elements is fundamentally not a numerical problem, but a structural one. The periodic table ultimately terminates not because a single physical interaction fails first, but because the overall structure can no longer sustain stable closure.

The mathematical formulation of this structural criterion will be developed in Chapter 4, while its scientific implications for chemical structures and the emergence of life will be explored in later chapters.

Concluding Statement:

The ultimate limit of the periodic table is determined not solely by nuclear forces or relativistic effects, but by the exhaustion of structural closure. Stable elements occupy only a finite structural window, and the termination of the periodic table marks the first irreversible breakdown of structural consistency.

Section 6. Why Must Hierarchical Structures Possess a Finite Stability Limit?

Abstract:

The termination of the periodic table raises a broader question: **Can stable structures continue to grow indefinitely?**

This question extends far beyond the chemical elements. Similar limitations appear throughout mathematics, physics, the life sciences, and complex systems. As hierarchical complexity increases, the number of interactions grows rapidly, feedback networks become increasingly intricate, and the range of stable configurations progressively narrows.

This book proposes that **finite stability is not a phenomenon unique to any single discipline, but a universal property shared by complex structures.**

Accordingly, the Structural Periodic Table is concerned not merely with where the periodic table ends, but with a more fundamental question: **Why does every stable hierarchy ultimately encounter its own structural boundary?**

I. From the Upper Limit of Elements to the Upper Limit of Hierarchies

The previous section discussed why stable chemical elements must possess an upper limit. A more fundamental question naturally follows: **Is this limitation unique to the elements?**

Many natural systems exhibit remarkably similar behavior:

- stable regions within the periodic table;
- islands of nuclear stability;
- synthesis limits for giant molecules;
- carrying capacities in ecological systems;
- complexity limits in information systems;
- undecidable regions in mathematical systems.

Although these phenomena belong to different disciplines, they all suggest the same structural principle: **Stable structures cannot increase indefinitely.**

II. Why Do Higher Hierarchical Levels Become Less Stable?

Increasing hierarchy does not merely introduce additional objects; it dramatically increases the number of relationships among them. As hierarchical complexity grows, at least three fundamental changes occur.

1) Increasing Degrees of Freedom

New structural components continually emerge, interactions multiply rapidly, and the system's state space expands accordingly.

2) Increasingly Complex Feedback Networks

New connections generate increasingly intricate feedback loops, allowing local perturbations to propagate more easily throughout the entire system.

3) Progressively Stronger Stability Requirements

Higher-level systems must simultaneously satisfy:

- local stability;
- global coordination;
- cross-scale consistency.

Consequently, the number of structures capable of satisfying all these requirements becomes progressively smaller, and the **finite structural window** continually narrows.

III. Stable Existence Requires Three Forms of Structural Closure

According to the Structural Periodic Table, every hierarchy capable of long-term existence must satisfy at least three forms of structural closure.

1) Local Closure

Local structural components remain internally stable.

2) Dynamic Closure

The system can continue evolving without rapidly collapsing.

3) Global Closure

Feedback remains coherent across different organizational scales, allowing the entire hierarchy to maintain long-term stability.

Only when all three forms of closure are simultaneously achieved does a hierarchy become a genuine **Existence Level.**

IV. Why Does the Stability Window Become Progressively Narrower?

As structural complexity continues to increase, local perturbations propagate more rapidly, feedback loops increasingly interfere with one another, and consistency across multiple scales becomes progressively more difficult to maintain.

Consequently, although many structures may exist in principle, only a relatively small subset can remain stable over long periods.

Stable regions therefore evolve from broad continuous domains into isolated stability windows and may eventually disappear altogether. Hierarchy may continue to increase, but structural stability does not.

V. Finite Stability Is a Universal Structural Principle

This phenomenon is not unique to the periodic table.

Comparable stability limits appear throughout science:

- rapid decay of superheavy nuclei;
- transitions to chaos in high-dimensional dynamical systems;
- cascading instabilities in complex networks;
- undecidable regions in mathematical logic;
- evolutionary constraints on biological complexity.

These diverse examples suggest that **finite stability is not the consequence of any single physical theory, but a universal property of complex hierarchical organization.**

VI. Implications for the Structural Periodic Table

If the Structural Periodic Table is valid, it must itself obey the same structural principles.

Accordingly, this framework does **not** propose that hierarchical organization can increase without limit. Instead, it argues that only those hierarchical levels capable of satisfying the three closure conditions possess genuine structural existence.

Levels that fail to achieve structural closure may remain useful theoretical extensions, but they cannot be regarded as stable existence levels. Thus, the existence of an upper stability limit is not a weakness of the theory; rather, it is one of the defining characteristics of a mature scientific framework.

VII. Implications for the Nine-Level Framework

The nine-level framework proposed in this book does not imply that every level possesses equal structural stability. On the contrary, as hierarchical complexity increases, the corresponding stability window may progressively shrink.

If future research demonstrates that higher organizational levels become increasingly difficult to stabilize, this should not be regarded as a defect of the Structural Periodic Table. Instead, it may simply reflect a universal property shared by complex systems.

Accordingly, what ultimately matters is not **how many hierarchical levels exist**, but **whether each level satisfies the conditions for structural closure.**

VIII. Summary

The periodic table teaches us that stable elements occupy only a finite region.

The Structural Periodic Table extends this principle by proposing that **every stable hierarchy also possesses a finite stability limit.**

Hierarchical complexity may continue to increase, but structural stability cannot increase without bound. Accordingly, every mature structural theory must explain not only how structures emerge, but also where structural stability ultimately breaks down and why.

Only a theory capable of identifying its own structural boundaries can genuinely claim scientific significance.

Concluding Statement:

The periodic table demonstrates that stable elements occupy only a finite region. The Structural Periodic Table extends this principle to all hierarchical systems: complexity may continue to increase, but structural stability cannot grow without bound. Every stable hierarchy exists within a finite structural window, and every mature theory must identify the boundaries of its own validity.

Keywords:

Structural unification; generative framework; interdisciplinary synthesis; structural dynamics; periodicity of knowledge; mathematical physics methodology.

References：（All of Chapter Two）

[1] **Einstein, A.** (1916). *The Foundation of the General Theory of Relativity*. Annalen der Physik.

[2] **Hilbert, D.** (1902). *The Foundations of Geometry*. Open Court.

[3] **Grothendieck, A.** (1984). *Esquisse d'un Programme*. Manuscript.

[4] **Weyl, H.** (1952). *Symmetry*. Princeton University Press.

[5] **Dirac, P. A. M.** (1930). *The Principles of Quantum Mechanics*. Oxford University Press.

[6] **Kuhn, T. S.** (1962). *The Structure of Scientific Revolutions*. University of Chicago Press.

[7] **Anderson, P. W.** (1972). More is Different. *Science*, 177(4047), 393–396.

[8] **Laughlin R. & Pines, D.** (2000). The Theory of Everything. *PNAS*, 97(1), 28–31.

[9] **Wigner, E.** (1960). The Unreasonable Effectiveness of Mathematics in the Natural Sciences. *Communications in Pure and Applied Mathematics*, 13(1), 1–14.

[10] **Penrose, R.** (2004). *The Road to Reality*. Jonathan Cape.

Chapter Three: Scientific Validation Framework and Methodology for the Structural Periodic Table

Figure 1-3: Crop Circle Illustrating a Galactic Civilization. See book《*Crop Circle*》for details.

Chapter Overview:

Any theory that seeks to enter scientific discourse must ultimately confront the same fundamental question: **Can it be tested?**

The Periodic Table was accepted by the scientific community not merely because it provided a successful classification of the chemical elements, but because it predicted previously unknown elements whose existence was subsequently confirmed experimentally.

Likewise, if the **Structural Periodic Table** is to develop beyond a philosophical concept into a scientific framework, it must provide clear and testable pathways for empirical evaluation.

This chapter does **not** attempt to demonstrate that the Structural Periodic Table has already been established. Instead, it asks a more fundamental question: **If the Structural Periodic Table represents a genuine structural principle of nature, how should it be tested through experiments, computation, and structural analysis?**

To address this question, the chapter proposes a four-level validation methodology.

1. Level One Validation: Chemical Reaction Networks

To examine whether predictable structural closure exists in the relationship between structural generation and reaction pathways.

2. Level Two Validation: Molecular Product Geometry

To investigate whether structural constraints possess predictive power over the stable geometrical configurations of molecular products.

3. Level Three Validation: Structural Reconstruction of Periodicity

To determine whether the Structural Periodic Table can reconstruct and reinterpret periodic behavior from structural principles rather than solely from electronic configurations.

4. Level Four Validation: An Integrated Validation Road-map

To integrate the preceding validation levels into a unified scientific strategy capable of guiding future theoretical, computational, and experimental investigations.

These four levels correspond to an ascending structural sequence: **Reaction → Molecular Geometry → Periodicity → Integrated Validation**

They also correspond to the hierarchical organization of the Structural Periodic Table: **Line | → Circle O → Hierarchical Structure → Scientific Integration**

If these directions receive consistent theoretical and experimental support, the Structural Periodic Table may become more than a structural language or philosophical interpretation. It may instead develop into a scientific framework possessing genuine explanatory and predictive capability.

Accordingly, the objective of this chapter is **not** to provide definitive proof, but to establish a practical pathway through which structural theory may progressively evolve into experimentally testable science.

With the rapid development of **AI-driven laboratories, high-throughput computation, autonomous experimentation, and digital research platforms**, structural theories are, for the first time, becoming amenable to large-scale, cross-model scientific validation.

Scope of This Chapter :

This chapter establishes the **scientific validation framework and methodology** for the Structural Periodic Table; it does **not** claim to complete its scientific verification.

The substantive validation of the proposed structural principles is developed in **Part II** of this book.

There, **mathematics, physics, and chemistry** respectively provide the first, second, and third layers of structural evidence. Together, these disciplines constitute the scientific foundation upon which the validity of the Structural Periodic Table may ultimately be evaluated.

Section 1. Scientific Validation Principles of the Structural Periodic Table

Abstract:

The central objective of validating the **Structural Periodic Table** is not to predict a particular numerical value. Rather, it seeks to determine:

- which classes of structures can exist as stable configurations,
- which structures cannot remain stable over long timescales,
- which structural pathways necessarily converge under multiple independent models.

Accordingly, the experimental validation of the Structural Periodic Table is not primarily concerned with measuring a physical constant. Instead, it focuses on testing three fundamental propositions:

1. **Can structural existence be systematically constrained?**
2. **Do multiple evolutionary pathways converge toward the same stable structures?**
3. **Does structural stability obey hierarchical criteria?**

I. Multi-Model Consistency Tests

Proposition: Do different experimental pathways inevitably generate the same class of stable structures?

This question can be investigated across diverse scientific systems, including: Quantum many-body systems; Chemical reaction networks; Turbulent fluid structures;Ecological evolutionary systems.

Different initial conditions and parameter pathways can be designed to examine whether the systems ultimately converge toward similar structural attractors.

If a broad range of systems converges toward equivalent structural organizations despite differing underlying models, this would support the **Fourth-Level Structural Consistency Principle** proposed by the Structural Periodic Table.

Conversely, if multiple fundamentally incompatible yet equally stable structures persist across independent models, the structural adjudication hypothesis would require reconsideration.

II. Structural Admissibility Tests

The Structural Periodic Table proposes that **not every theoretically conceivable structure is physically admissible**.

This proposition can be examined by constructing extreme theoretical models, such as:

- highly symmetric high-dimensional systems;
- systems without closure constraints;
- systems lacking hierarchical differentiation.

The key questions are whether such systems can:

- remain stable over extended periods;
- reproduce consistently;
- exist robustly across different modeling frameworks.

If these extreme structures prove to be generally stable, the structural admissibility criterion would be challenged.

III. Generativity Tests

A central hypothesis of the Structural Periodic Table is that the stable emergence of complex systems is governed by structural principles rather than by chance alone.

Possible experimental platforms include:

- artificial chemical reaction networks;
- self-organizing materials;
- quantum many-body simulations;
- AI-driven generative models.

The primary questions are:

- Do structural "generation dead zones" exist?
- Are certain classes of structures fundamentally unable to emerge as stable configurations?

If stable structural generation is found to occur only within identifiable structural boundaries, this would provide evidence supporting the proposed structural framework.

IV. Cross-Scale Consistency Tests

The Structural Periodic Table predicts that local structural rules must remain compatible with global structural stability.

This hypothesis can be examined by investigating whether:

- microscopic fluctuations ultimately evolve toward stable macroscopic organization;
- macroscopic structural stability constrains microscopic degrees of freedom.

If long-term stable systems exhibiting persistent cross-scale inconsistency are identified, the hierarchical assumptions underlying the Structural Periodic Table would need to be re-evaluated.

V. Predictive Validation

A scientific structural framework should possess predictive capability. If the Structural Periodic Table can successfully predict:

- the emergence of previously unknown structural classes;
- theoretical frameworks likely to fail under structural criteria;
- new forms of interdisciplinary structural unification;

and these predictions are subsequently confirmed, then the Structural Periodic Table advances beyond an interpretive framework toward a genuinely predictive scientific theory.

VI. Differences from Traditional Experimental Science

Traditional experimental science primarily asks: **Do the measured numerical results agree with theoretical predictions?**

The Structural Periodic Table instead asks: **Is a given structure fundamentally admissible?**

These represent two complementary but distinct levels of scientific inquiry. The Structural Periodic Table is not intended to replace conventional experimentation. Rather, it provides structural admissibility criteria that help define the boundaries within which experimental phenomena can occur.

VII. The Fundamental Principle of Scientific Validation

The Structural Periodic Table is, in principle, falsifiable.

Its central hypothesis would be challenged by the discovery of a system that is simultaneously:

- stable,
- reproducible,
- consistent across multiple independent models,
- and yet cannot be embedded within any structural hierarchy proposed by the framework.

Such a finding would require the Structural Periodic Table to be revised or abandoned.

A scientific theory must specify not only what it explains, but also how it can be tested. By proposing explicit experimental strategies, the Structural Periodic Table aims to move beyond conceptual speculation toward a scientifically testable research program.

Concluding Statement: The Structural Periodic Table is not intended to demonstrate that a particular equation is correct. Its purpose is to provide structural criteria for determining which classes of equations and theoretical frameworks are capable of representing stable structures in nature.

Section 2. First-Level Validation – Chemical Reaction Networks

Abstract:

A fundamental experimental question is whether **all possible chemical combinations are capable of generating complex structures**.

The Structural Periodic Table proposes that the answer is **no**.

Only reaction networks satisfying three structural adjudication criteria are expected to develop into higher levels of structural organization. Conversely, if experiments demonstrate that arbitrary random reaction systems can generate increasingly complex structures without structural constraints, the central hypothesis of the Structural Periodic Table would be challenged.

I. Why Choose Chemical Reaction Networks as the First Validation Platform?

If the Structural Periodic Table is valid, then:

- chemical reactions do not occur without structural constraints;
- reaction networks do not evolve arbitrarily;
- structural generativity is not unlimited.

Instead, their evolution should satisfy three fundamental structural criteria:

1) Structural Consistency
2) Structural Stability
3) Structural Generativity

Chemical reaction networks provide an ideal validation platform because they are:

- experimentally controllable;
- highly reproducible;

- directly observable;
- quantitatively measurable.

If the Structural Periodic Table possesses genuine predictive capability, it should first be able to distinguish reaction networks that spontaneously develop stable structures from those that inevitably collapse.

II. General Experimental Framework

2.1 Objective

To test the following hypothesis: **Only reaction networks satisfying the three structural adjudication criteria can develop stable, reproducible, and scalable structural hierarchies.**

2.2 Experimental Systems

Three representative classes of reaction systems may be compared:

1) Random reaction mixtures (without structural design)
2) Path-constrained reaction networks (artificially restricted pathways)
3) Autocatalytic feedback systems (possessing structural generative potential)

Representative examples include:

- simple organic reaction systems involving formic acid, acetaldehyde, ammonia, and related compounds;
- the classical **Belousov–Zhabotinsky oscillating reaction**;
- simple autocatalytic molecular reaction networks.

III. Experimental Criteria for Structural Adjudication

3.1 Criterion One: Structural Consistency

The following questions are examined:

• Under varying experimental conditions (temperature, concentration perturbations, etc.),
• does the reaction network consistently converge toward the same structural outcome?

If the resulting structures are highly sensitive to experimental conditions and lack reproducibility, the consistency criterion is not satisfied.

3.2 Criterion Two: Structural Stability

The following properties are evaluated:

- the existence of stable concentration distributions;
- periodic oscillations or stable attractors;
- long-term persistence of structural organization.

Rapid collapse or unbounded divergence indicates failure of the stability criterion.

3.3 Criterion Three: Structural Generativity

The experiment examines whether the system produces:

- new structural hierarchies;
- new classes of molecular products;
- increasing network complexity.

If the system merely consumes reactants repeatedly without generating new structural organization, the generativity criterion is not fulfilled.

IV. Predictions of the Structural Periodic Table

If the Structural Periodic Table is correct, it predicts:

1) Random reaction mixtures will not develop stable generative networks.

2) Path-constrained systems may remain temporarily stable but will not generate higher structural levels.

3) Feedback-closed reaction systems will evolve into stable and expandable structural organizations.

In other words, **Structural generativity is expected to emerge only within reaction networks possessing structural closure.**

V. Quantitative Evaluation

To enable objective comparisons, several measurable indicators may be introduced.

5.1 Network Complexity Index

Possible measures include:

- number of nodes;
- edge connectivity;
- number of feedback loops.

5.2 Stable Attractor Analysis

Possible indicators include:

- convergence of time-series trajectories;
- stability of oscillation periods.

5.3 Structural Hierarchy Transition Index

Representative measures include:

- number of newly generated molecular classes;
- changes in molecular complexity distributions.

Agreement between these quantitative indicators and the predictions of the Structural Periodic Table would provide evidence supporting the proposed structural adjudication framework.

Conversely, the absence of reproducible structural patterns would argue against the theory.

VI. Extension to AI-Driven Autonomous Laboratories

The proposed validation framework can naturally be extended through modern automated research platforms, including:

- robotic high-throughput experimentation;
- AI-guided reaction-pathway exploration;
- automated reaction-network modeling.

AI systems may be employed to:

1) generate reaction networks automatically;
2) simulate their structural evolution;
3) identify networks satisfying the three structural adjudication criteria.

If generative reaction networks consistently appear only within structurally closed regions of the search space, the Structural Periodic Table would receive substantial empirical support.

VII. Statement of Falsifiability

This validation framework is explicitly falsifiable. If experimental evidence demonstrates that:

- structural generativity is unrelated to structural stability;
- reaction-network behavior is entirely random;
- no predictive structural patterns can be identified;

then the Structural Periodic Table must be revised or abandoned.

VIII. Ultimate Criterion

If the Structural Periodic Table cannot predict which classes of reaction networks will develop self-organized structures, it cannot be regarded as a scientifically useful structural framework.

Concluding Statement:

A theory that cannot determine the structural conditions for generativity remains philosophy rather than science. Likewise, if the Structural Periodic Table cannot predict the emergence of self-organization in chemical reaction networks, it must ultimately be rejected.

Section 3. Second-Level Validation — Molecular Product Geometry

Abstract:

The molecular geometry of a chemical product is proposed not to be merely the accidental consequence of electron arrangements, but the outcome of structural closure governed by underlying structural principles.

If molecular geometries are found to belong to a finite set of stable structural classes, this would provide first-principles support for the Structural Periodic Table.

I. Why Choose Molecular Geometry as the Second Validation Platform?

If the Structural Periodic Table is valid, then molecular geometry should not be viewed simply as:

- an empirical observation;
- an accidental consequence of electron distributions;
- or merely a balance of local electronic interactions.

Instead, molecular geometry should represent the structural outcome of a closure process governed by fundamental organizational principles.

Molecular geometry is an ideal validation platform because it possesses three important characteristics:

1) It can be measured precisely using techniques such as X-ray crystallography and spectroscopy.

2) Different molecular geometries can be directly compared (linear, bent, tetrahedral, etc.).

3) Similar structural principles can be examined across many different chemical systems.

If the Structural Periodic Table possesses predictive capability, it should be able to anticipate classes of molecular geometry without relying exclusively on conventional frameworks such as **VSEPR theory** or **molecular orbital theory**.

II. Structural Prediction Hypothesis

The Structural Periodic Table proposes the following hypothesis: **Molecular geometry is primarily the result of structural closure rather than solely the consequence of local electron-repulsion equilibrium.**

Accordingly, familiar molecular geometries — including

- linear,
- bent,
- trigonal pyramidal,
- tetrahedral,

may be interpreted as different structural solutions satisfying distinct closure conditions instead of merely minimizing local electronic repulsion.

III. Three Structural Criteria for Molecular Geometry

A stable molecular geometry should satisfy three structural criteria.

3.1 Criterion One: Connectivity Closure

Atoms must form a complete connectivity network without incomplete bonding pathways.

3.2 Criterion Two: Tension Minimization

The spatial arrangement should minimize overall structural tension by satisfying:

- uniform geometric constraints;
- minimal directional conflict;
- balanced angular distribution.

3.3 Criterion Three: Structural Identity

A molecular geometry should remain structurally stable under:

- small perturbations;
- thermal fluctuations;
- changes of physical phase.

Failure to satisfy these three criteria indicates that the geometry cannot remain structurally stable over time.

IV. Illustrative Structural Interpretations

4.1 Why is CO_2 Linear?

From a structural perspective:

- the two terminal atoms are symmetrically arranged;
- the central atom acts as a single connecting node;
- no directional conflict is introduced.

A bent geometry would:

- reduce symmetry;
- increase structural tension;
- introduce unnecessary geometric complexity.

Therefore, the linear configuration represents the simplest structurally closed arrangement.

4.2 Why is H_2O Bent?

Structurally,

- the central atom connects two distinct directions;
- spatial directional constraints cannot be neglected;
- forcing a linear geometry would prevent stable spatial closure.

Consequently, the bent configuration represents the minimum-tension closed structure.

4.3 Why is NH_3 Trigonal Pyramidal?

Structurally,

- three bonding directions must be accommodated;
- a planar configuration introduces directional conflicts;
- slight three-dimensional displacement reduces structural tension.

Thus, the trigonal pyramidal geometry provides a stable structural closure.

4.4 Why is CH_4 Tetrahedral?

Structurally,

- four equivalent bonding directions must be distributed uniformly;
- planar arrangements inevitably introduce conflict;
- three-dimensional symmetry provides the only uniformly closed configuration.

Accordingly, the tetrahedral geometry represents the minimum-tension structural solution.

V. The Central Validation Question

The primary prediction of the Structural Periodic Table is not a specific bond angle, but rather that **the number of stable molecular geometry classes is finite.**

If experiments were to demonstrate

- infinitely many stable random geometries, or
- completely unconstrained angular distributions,

then the structural adjudication hypothesis would be challenged.

Current observations instead indicate that molecular geometries belong to a limited collection of recurring structural classes, including:

- linear;
- planar;
- trigonal pyramidal;
- tetrahedral;
- octahedral;

and several other well-defined configurations.

Such observations are consistent with the structural closure hypothesis.

VI. Experimental Validation Framework

6.1 Method One: AI-Based Geometry Generation

An AI-based validation framework may proceed by:

1) generating random atomic connectivity graphs;
2) optimizing molecular geometries without explicitly applying VSEPR rules;
3) statistically analyzing the resulting stable geometry classes.

If the optimized geometries consistently converge toward a finite number of structural classes, this would support the Structural Periodic Table.

6.2 Method Two: Prediction of Unknown Molecular Geometries

Molecules whose geometries have not yet been experimentally determined may first be predicted using only structural adjudication principles and then compared with subsequent experimental measurements.

Prediction accuracy significantly exceeding random expectation would provide evidence supporting the structural framework.

VII. Comparison with Conventional Chemical Theories

Traditional chemical theories primarily address the question: **Why does this molecule adopt a particular bond angle?**

The Structural Periodic Table instead asks: **Why do stable molecular geometries belong to only a limited number of structural classes?**

Its primary concern is therefore the finiteness and organization of molecular geometry classes rather than the precise numerical value of individual bond angles.

VIII. Statement of Falsifiability

The proposed framework is falsifiable. If future experiments demonstrate:

- an unlimited proliferation of stable molecular geometry classes;
- no concentration toward preferred structural configurations;
- no observable tendency toward structural tension minimization;

then the Structural Periodic Table must be revised or abandoned.

Concluding Statement: The Structural Periodic Table does not seek to predict every bond angle. Rather, it seeks to explain why stable molecular geometries are confined to a finite number of structural classes.

Accordingly, if the Structural Periodic Table cannot explain this finite structural organization, its proposed structural principles cannot adequately account for the hierarchical organization of chemical structure.

Section 4. Third-Level Validation – Structural Reconstruction of Chemical Periodicity

Abstract:

The ultimate validation of the Structural Periodic Table lies in determining whether it can account for the fundamental characteristics of the Periodic Table itself, including:

- the increasing length of successive periods;
- the symmetry of elemental groups;
- the finite number of stable elements; and
- the necessary emergence of noble gases.

If these structural features cannot be reconstructed from structural principles, the Structural Periodic Table cannot be regarded as a valid scientific framework.

I. The Fundamental Question

Conventional chemistry explains periodicity through electronic structure:

- periodicity arises from electron-shell filling;
- quantum numbers determine orbital configurations;
- orbital configurations determine periodic chemical properties.

This establishes a chain of explanation extending from microscopic quantum mechanics to macroscopic chemical behavior.

The Structural Periodic Table asks a more fundamental question: **Why should electron shells themselves exhibit hierarchical periodic organization?**

Why do elements appear in discrete hierarchical levels rather than forming a continuous spectrum?

The essential issue is therefore whether periodicity is merely a consequence of electronic theory or a necessary outcome of structural closure.

II. Structural Hypothesis

Rather than beginning with electronic wave equations, the Structural Periodic Table begins with three structural axioms:

1) Point (•): structural identity must form through local closure.
2) Line (|): structural evolution requires directional organization.
3) Circle (O): stable systems require global closure.

Since chemical elements belong to the **Point (•) hierarchy**, their existence must satisfy three conditions:

- local structural closure;
- hierarchical differentiation;
- stable reproducibility.

If elemental properties varied continuously without structural boundaries, the concept of discrete chemical elements would itself lose meaning.

III. The Structural Necessity of Periodicity

Within the Structural Periodic Table, periodicity is interpreted not primarily as the consequence of orbital filling, but as the structural requirement that identities emerge through successive hierarchical closures.

3.1 Principle of Hierarchical Saturation

Once a structural level reaches its stable capacity, the system must transition to the next hierarchical level.

This is formally analogous to electron shells becoming filled before higher shells begin to populate.

The structural interpretation, however, is that every stable hierarchy possesses a finite closure capacity.

3.2 Principle of Structural Differentiation

If elemental properties changed continuously, distinct elemental families would not exist. Instead, nature exhibits well-defined structural groupings, including:

- alkali metals;
- halogens;
- noble gases.

These recurring families suggest an underlying structural symmetry rather than arbitrary chemical classification.

3.3 Principle of Finite Structural Existence

If elements remained indefinitely stable as atomic number increased, the Periodic Table would extend without limit. Instead, heavy elements become progressively unstable.

This suggests that structural closure itself possesses an upper limit.

IV. A Structural Reconstruction of Periodicity

Instead of beginning with the Schrödinger equation, the Structural Periodic Table proposes a structural reconstruction based on hierarchical criteria.

4.1 Step One: Determining Structural Capacity

Each structural level must satisfy three requirements:

- finite occupancy;
- balanced directional organization;
- avoidance of excessive internal structural tension.

These constraints naturally produce increasingly larger hierarchical levels.

4.2 Step Two: Emergence of Structural Symmetry

When a structural level reaches completion, it enters a state of maximum symmetry. This corresponds naturally to the noble gases.

The opening of the next structural level returns the system to a new stage of structural instability before another stable configuration is reached.

4.3 Step Three: Structural Upper Bound

When structural tension within the coupled nuclear-electronic system exceeds the available closure capacity, stable elemental identities can no longer be maintained.

This provides a structural interpretation for:

- the radioactivity of heavy elements;
- the instability of superheavy elements.

V. Testable Predictions

The Structural Periodic Table makes several predictions that are, in principle, experimentally testable.

1) Prediction One: A Finite Number of Stable Elements

The existence of an upper structural limit is a necessary consequence of structural closure rather than an accidental limitation of current experiments.

2) Prediction Two: Persistent Hierarchical Organization

Even under extreme physical conditions, structural differentiation into elemental families should continue to emerge.

3) Prediction Three: Future Elements Should Preserve Hierarchical Organization

If additional stable electron shells are discovered, they should correspond to new structural closure levels rather than arbitrary extensions of the existing Periodic Table.

VI. Relationship to Electronic Theory

The Structural Periodic Table does not reject electronic theory. Instead, it proposes that electronic structure represents one specific physical realization of more general structural closure principles.

Electronic orbitals describe the **mechanism**, whereas structural closure provides the **structural criterion** governing stable organization.

VII. Statement of Falsifiability

The Structural Periodic Table would require revision or abandonment if future evidence demonstrates that:

- stable elements can extend indefinitely;
- periodic organization disappears at sufficiently high atomic numbers while structural stability remains;
- elemental properties become continuously distributed without recognizable structural differentiation.

VIII. Concluding Statement

The Structural Periodic Table does not seek merely to reinterpret electron-shell filling. Rather, it seeks to explain why periodicity itself emerges as a consequence of hierarchical structural closure. If periodic organization were to disappear entirely, the concept of discrete chemical elements would likewise lose its structural foundation.

Section 5. An Integrated Validation Roadmap for the Structural Periodic Table

Abstract:

Every scientific theory must ultimately withstand systematic verification. The previous three sections introduced three complementary validation levels:

- chemical reaction networks;
- molecular geometries;
- chemical periodicity.

These are not independent experiments, but successive stages of a unified validation pathway extending from local generation, through structural closure, to hierarchical periodicity.

If evidence from experiments, computational simulations, and artificial intelligence consistently supports all three levels, the Structural Periodic Table may evolve beyond a structural interpretation framework into a predictive scientific framework.

I. Why Is Multi-Level Validation Necessary?

No comprehensive scientific theory can be established through a single experiment alone. There are several reasons for this:

- a single experiment verifies only a local phenomenon;
- different theories may explain the same experimental observations;
- a genuine structural theory must remain consistent across multiple hierarchical levels simultaneously.

For this reason, the Structural Periodic Table proposes a strategy of **Hierarchical Consistency Verification.**

Rather than testing isolated phenomena, it examines whether the same structural principles operate consistently across different organizational scales.

II. The Three-Level Validation Framework

2.1 First Level: Generation Verification

- **Object of study:** Chemical reaction networks.
- **Central question:** Can only structurally closed reaction networks continuously generate new stable structures?
- **Primary criterion:** Generativity
- **Corresponding structural role:** Line (|)

2.2 Second Level: Structural Closure Verification

- **Object of study:** Molecular geometries.
- **Central question:** Do stable molecular structures belong to a finite set of structural classes?
- **Primary criterion:** Structural Closure
- **Corresponding structural role:** Circle (O)

2.3 Third Level: Hierarchical Periodicity Verification

- **Object of study:** Chemical periodicity.
- **Central question:** Can periodic behavior be reconstructed from structural principles rather than solely from electronic mechanisms?
- **Primary criterion:** Hierarchical Periodicity
- **Corresponding structural role:** Point (•)

III. The Validation Roadmap

The complete validation pathway may therefore be summarized as: **Reaction Networks → Generation → Molecular Geometry → Structural Closure → Chemical Periodicity → Hierarchical Organization → Structural Periodic Table**

Accordingly, the Structural Periodic Table is not founded upon a single experiment. Its validity depends upon the consistency of structural principles across multiple hierarchical levels.

IV. Artificial Intelligence and Future Validation

Recent advances in artificial intelligence, high-throughput computation, and automated laboratories make large-scale structural validation increasingly feasible. Future AI systems may be able to:

- automatically generate reaction networks;
- search for stable molecular geometries;
- identify periodic structural patterns;
- discover structural vacancies;
- compare the structural consistency of competing scientific theories.

Artificial intelligence therefore represents more than a computational tool. It may become one of the principal experimental platforms for validating the Structural Periodic Table.

V. The Ultimate Criterion of Falsifiability

The Structural Periodic Table does **not** require that every experiment immediately confirms its predictions. Instead, it proposes explicit conditions under which the theory should be revised or abandoned.

For example, the framework would be seriously challenged if future evidence demonstrated that:

- arbitrary reaction networks can indefinitely generate stable complex structures;
- stable molecular geometries do not belong to a finite set of structural classes;
- chemical periodicity cannot be reconstructed from any coherent structural principle.

Consequently, the Structural Periodic Table should be regarded as a **scientific hypothesis** that is explicitly open to experimental testing rather than as an unfalsifiable philosophical proposition.

VI. Concluding Remarks

The previous sections validated three successive structural principles:

- **Generation**
- **Closure**
- **Periodicity**

This final section addresses a more fundamental question: **Do these three principles together represent a single underlying structural law?**

If the answer is affirmative, the Structural Periodic Table represents far more than a new method of scientific classification.

It becomes a unified structural framework capable of connecting mathematics, physics, chemistry, and ultimately the emergence of living systems through a common structural language.

Concluding Statement: A single experiment may verify a single phenomenon; a structural principle can be validated only through consistency across multiple hierarchical levels. The Structural Periodic Table is therefore tested not by any individual experiment, but by whether nature itself consistently follows a common structural language.

References (Chapter 3):

[1] Popper, K. R. (1959). *The Logic of Scientific Discovery*. Hutchinson.

[2] Kuhn, T. S. (1962). *The Structure of Scientific Revolutions*. University of Chicago Press.

[3] von Bertalanffy, L. (1968). *General System Theory: Foundations, Development, Applications*. George Braziller.

[4] Prigogine, I., & Stengers, I. (1984). *Order Out of Chaos: Man's New Dialogue with Nature*. Bantam Books.

[5] Barabási, A.-L. (2016). *Network Science*. Cambridge University Press.

[6] Mitchell, M. (2009). *Complexity: A Guided Tour*. Oxford University Press.

[7] Lehn, J.-M. (1995). *Supramolecular Chemistry: Concepts and Perspectives*. Wiley-VCH.

[8] Whitesides, G. M., & Ismagilov, R. F. (1999). "Complexity in Chemistry." *Science*, 284(5411), 89–92.

[9] Goodfellow, I., Bengio, Y., & Courville, A. (2016). *Deep Learning*. MIT Press.

[10] Jumper, J., Evans, R., Pritzel, A., et al. (2021). "Highly Accurate Protein Structure Prediction with AlphaFold." *Nature*, 596, 583–589.

[11] Chang, J. (2026). *Recursive Self-Organizing Evolutionary Breakthrough: Unifying Natural, Social and Life Sciences*. Universal Publishing.

[12] Chang, J. (2013). *Grand Ultimate Theory*. Universal Publishing.

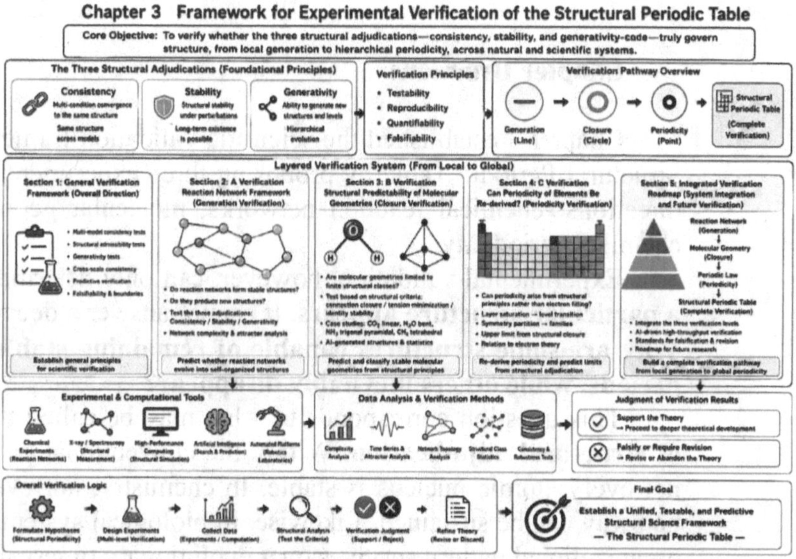

Figure 1-3-5-1: Experimental Validation Framework of the Structural Periodic Table

Chapter Four: The Structural Adjudication Framework of the Structural Periodic Table

Figure 1-4: Crop Circle Illustrating a Galactic Civilization , For details, see book 《*Crop Circle*》.

Chapter Overview:

Chapter 3 established the scientific validation framework of the Structural Periodic Table by proposing three experimentally testable directions: chemical reaction networks, molecular geometries, and chemical periodicity.

Experimental validation, however, can only determine **whether a particular structure appears**. It cannot answer a deeper question: **Why are some structures capable of remaining stable over long periods, while others inevitably disappear?**

This question corresponds to what may be called the **problem of structural adjudication**. Within the conventional periodic table, not every atomic nucleus is stable. In chemistry, not every reaction pathway can be sustained. Likewise, in biological systems, not every organizational pattern survives over evolutionary timescales.

Generation alone is therefore insufficient to explain the natural world. There must also exist a mechanism by which structures are continuously filtered according to their stability.

This chapter proposes a **Structural Adjudication Framework** for the Structural Periodic Table.

Its central idea is straightforward: **Structures are not merely generated — they must also undergo structural adjudication.**

Here, *adjudication* does not refer to an external decision imposed upon a system. Rather, it denotes the intrinsic process by which a structure continuously evaluates its own capacity for stable existence during its evolution. Only those structures satisfying appropriate stability conditions are capable of progressing to higher hierarchical levels and maintaining long-term existence.

To formalize this idea, the chapter first introduces a mathematical representation of hierarchical structural stability and then develops the framework through four successive levels.

Level I – Generativity

The generative criterion for chemical reaction networks.

This level investigates how the number of reaction pathways influences structural stability and the emergence of new organizational levels.

Level II – Structural Consistency

The multi-scale consistency criterion for molecular geometries.

Here the focus is the coordination between local bonding constraints and global geometric stability.

Level III – Feedback Stabilization

The feedback fixed-point structure underlying chemical periodicity. This section explains why the periodic table naturally develops stable and repeatable hierarchical organization.

Level IV – Hierarchical Boundary

The finite repetition of elemental families.

The final level explores why periodic structures possess intrinsic upper limits and under what conditions hierarchical organization ultimately terminates.

Together, these four levels represent the complete adjudication pathway of the Structural Periodic Table: **Generation → Coordination → Stabilization → Boundary**

This sequence describes the transition from the emergence of structure to its long-term persistence. Accordingly, the purpose of this chapter is **not** to construct an entirely new mathematical formalism. Instead, it seeks to answer a more fundamental question:

Once a structure is allowed to emerge, what determines how long it may persist, how far it can develop, and where its hierarchical evolution must ultimately end?

Answering this question is essential if the Structural Periodic Table is to progress from a descriptive framework toward a genuinely predictive scientific framework.

From Validation to Adjudication

In simple terms, **Chapter 3 asks: How can the Structural Periodic Table be scientifically validated?**

Chapter 4 asks: Even after validation, how can we determine whether a structure is genuinely stable?

From a more formal perspective, the Structural Periodic Table requires its own **mathematical language of structural adjudication**.

This chapter therefore represents the conceptual transition: **Validation → Structural Adjudication**

Chapter 3	Chapter 4
Can the theory be validated?	Can the structure be adjudicated?
Experimental framework	Mathematical framework
Structural emergence	Structural selection
Reaction → Geometry → Periodicity	Generation → Stability → Boundary

Table 1-4-1: Comparison Between Scientific Validation (Chapter 3) and Structural Adjudication (Chapter 4)

Validation tells us whether a structure exists. Structural adjudication determines whether it deserves to persist.

Section 1. Why Does the Structural Periodic Table Require Structural Adjudication?

Abstract:

Every scientific theory requires a principle capable of distinguishing **what is allowed to exist** from **what cannot persist**.

The Periodic Table of Elements became one of the foundations of modern science not merely because it classifies chemical elements, but because it also distinguishes between elements that can exist stably, those that survive only briefly, and those that cannot form stable structures at all.

Likewise, if the Structural Periodic Table is to become a genuine scientific theory, it cannot remain a framework for structural classification alone.

It must also answer a more fundamental question: **Why are some structures capable of forming stable hierarchical organizations, whereas others inevitably collapse or fail to persist?**

To address this question, this chapter introduces the concept of **Structural Adjudication**. Structural adjudication is **not** the process of determining whether a scientific theory is correct.

Rather, it is the process of determining whether a structure possesses the capacity for sustained existence, continued evolution, and hierarchical development.

I. From Classification to Adjudication

The development of science generally proceeds through two stages. The first stage establishes systems of classification.

For example:

- biology classifies living organisms;
- chemistry classifies chemical elements;
- astronomy classifies stars and galaxies.

Classification answers the question: **"To which category does it belong?"**

The second stage asks a deeper question: **"Why is it able to exist?"**

A mature scientific theory does more than organize knowledge. It explains the origin of structural stability and predicts the limits within which structures may exist. The Structural Periodic Table must therefore evolve from **classification** toward **adjudication**.

II. Why Is Classification Alone Insufficient?

Classification by itself cannot explain many fundamental phenomena observed in nature. For example:

- Why do some elements remain stable while others decay rapidly?
- Why do certain chemical reactions continuously generate complex structures whereas others terminate quickly?
- Why do some scientific theories continue to develop while others are eventually abandoned?

These questions extend beyond classification itself.

They all point toward a deeper issue: **Which structures truly possess the qualifications for existence?**

Accordingly, the Structural Periodic Table must not only describe structures—it must also evaluate them.

III. What Is Structural Adjudication?

This book defines **Structural Adjudication** as **the evaluation of whether a structure can continue to generate, remain stable, and evolve toward higher hierarchical levels according to a unified set of structural principles.**

Structural adjudication is not concerned with

- what an object is called;
- which scientific discipline a theory belongs to.

Instead, it asks whether the structure

- can achieve stable closure;
- can maintain hierarchical consistency;
- can continue generating new organizational levels.

Structural adjudication therefore investigates the **conditions required for structural existence.**

IV. Why Must the Structural Periodic Table Possess Adjudicative Power?

A theory that merely classifies remains descriptive.
The greatness of the Periodic Table lies not simply in arranging elements but in predicting

- which elements had yet to be discovered;
- which elements could remain stable;
- which elements could never exist for extended periods.

Similarly, if the Structural Periodic Table is to become a predictive scientific framework, it must determine

- which structures are capable of generating new hierarchical levels;
- which structures remain confined to local organization;
- which structures ultimately lose stability because of internal inconsistency.

Adjudicative capability therefore distinguishes the Structural Periodic Table from an ordinary classification system.

V. Three Fundamental Objects of Structural Adjudication

In this book, structural adjudication focuses primarily on three fundamental properties.

1) Generativity

Can a structure continuously generate new forms of organization?

2) Stability

Can the structure maintain long-term internal consistency?

3) Hierarchy

Can the structure successfully evolve toward higher levels of organization?

Only structures satisfying all three conditions possess the potential for sustained development.

VI. From Scientific Validation to Mathematical Adjudication

The previous chapter established the scientific validation framework of the Structural Periodic Table. Validation addresses the question: **How can a theory be tested?**
The present chapter advances to a different question: **How can a structure itself be evaluated?**
Accordingly, this chapter does not focus on experimental procedures. Instead, it develops a mathematical language capable of describing

- structural generation;
- structural stability;
- structural feedback;
- hierarchical boundaries.

This mathematical language will serve as the common foundation for the subsequent discussions of mathematics, physics, chemistry, and ultimately living systems.

VII. Objectives of This Chapter

The objective of this chapter is not to construct a mathematical model for a particular scientific discipline. Instead, it seeks to establish a unified adjudication framework applicable across multiple disciplines.

The following sections will therefore discuss:

- why structural science requires a new mathematical language;
- how a unified structural evolution equation can be established;
- how four fundamental adjudication criteria — generation, closure, feedback, and boundary — can be formulated.

Together these constitute the mathematical foundation of the Structural Periodic Table.

VIII. Concluding Remarks

Classification determines **where** a structure belongs.

Structural adjudication determines **whether** it can endure.

The Periodic Table answers the question: **"How are the elements organized?"**

The Structural Periodic Table asks a more fundamental question: **"Which structures are capable of stable existence?"**

Concluding Statement:

Classification organizes knowledge. Structural adjudication determines existence. The Structural Periodic Table not only arranges structures — it evaluates which structures possess the capacity for sustained existence and continued evolution.

Section 2. Why Does the Structural Periodic Table Require a New Mathematical Language?

Abstract:

Mathematics has always been the fundamental language of science. From classical geometry to modern differential equations, from linear algebra to probability theory, successive mathematical frameworks have continually expanded humanity's ability to describe nature.

As scientific inquiry advances from relatively simple systems toward highly complex systems, however, describing quantities, spatial relationships, and dynamical evolution alone is no longer sufficient. More fundamental questions arise:

- Why do new structures emerge?
- Why are some structures capable of remaining stable?
- Why does structural growth eventually reach its limits?

The Structural Periodic Table investigates not individual objects but the common principles governing structural generation across multiple scientific disciplines.

Accordingly, this book proposes that describing structural periodicity requires a new mathematical language centered on four fundamental processes: **Generation, Closure, Feedback, and Boundary.**

I. What Does Traditional Mathematics Describe?

Modern mathematics has established an extraordinarily rich and rigorous theoretical system. For example,

- algebra investigates operational relationships;
- geometry studies spatial organization;
- calculus describes continuous change;
- probability theory analyzes stochastic behavior;
- topology examines continuous deformation.

These mathematical disciplines have profoundly advanced modern science. Nevertheless, they primarily address questions such as:

- How do quantities change?
- How are spatial relationships organized?
- How do systems evolve over time?

They do not directly explain:

- Why do entirely new structures emerge?
- Why do structures organize into stable hierarchical levels?
- Why do complex systems ultimately possess finite limits?

These belong to a higher level of structural inquiry.

II. What Does the Structural Periodic Table Study?

The Structural Periodic Table investigates

- how structures are generated;
- how they become stable;
- how they maintain themselves through feedback;
- and how they evolve toward new hierarchical levels.

Its object of study is therefore not

- a particular equation;
- a specific physical model;
- or an individual chemical system.

Instead, it seeks common structural evolution across diverse scientific disciplines. Examples include:

- the formation of mathematical theories;
- the emergence of physical laws;
- the organization of chemical matter;
- the evolution of living systems.

Although these subjects differ substantially, they all exhibit remarkably similar hierarchical structural behavior.

A unified structural language capable of transcending disciplinary boundaries therefore becomes necessary.

III. Why Is Traditional Mathematics Not Yet Sufficient?

Contemporary mathematics successfully describes an enormous range of complex phenomena. Nevertheless, several fundamental challenges remain concerning structural emergence itself.

1) The Emergence Problem

Most mathematical models begin with an existing state and describe its subsequent evolution. Relatively few explain why an entirely new organizational structure should emerge in the first place.

2) The Stability Problem

Different scientific disciplines have developed their own concepts of stability, including

- Lyapunov stability;
- thermodynamic stability;
- chemical equilibrium;
- evolutionarily stable strategies.

Despite these achievements, no unified stability criterion presently exists that applies consistently across multiple scientific disciplines.

3) The Boundary Problem

Many important questions remain without a common explanation. For example:

- Why is the number of stable chemical elements finite?
- Why is biological complexity limited?

• Why do mature theoretical systems eventually approach structural equilibrium?

These issues are usually studied independently. The Structural Periodic Table seeks a unified structural interpretation.

Traditional mathematics primarily describes: **Dynamics.**

Structural mathematics further investigates: **Existence.**

IV. What Should a New Mathematical Language Describe?

If the Structural Periodic Table is valid, its mathematical language should at least provide a unified description of four fundamental structural processes.

- **Generation:** How does a new structure emerge?
- **Closure:** How does a structure become an integrated and stable whole?
- **Feedback:** How does a stable structure continuously maintain itself?
- **Boundary:** Why does structural growth eventually terminate and transition toward a new hierarchical level?

Together these four processes constitute the fundamental cycle of structural evolution.

V. The Fundamental Philosophy of Structural Mathematics

This book proposes that mathematics should investigate not only quantitative relationships but also **structural roles.**

Traditional mathematics typically studies

- points;
- curves;
- surfaces;
- spaces.

Structural mathematics extends this perspective by introducing three fundamental structural roles:

- **Point** — irreducible existence;
- **Line** — directional generation;
- **Circle** — stable closure.

More importantly, it studies the continuous interactions among these structural roles through

- generation;
- feedback;
- and hierarchical transition.

Accordingly, the primary mathematical objects considered in this book are no longer merely geometric entities but **Structural Roles**.

VI. From Mathematical Language to Structural Language

The proposed mathematical language is not intended to replace existing mathematics. Rather, it is built upon the foundations of modern mathematics while introducing an additional level of structural organization.

Traditional mathematics asks: **How does a system evolve?**

Structural mathematics asks a more fundamental question: **Why is a system capable of existing at all?**

The former investigates the laws of dynamics.

The latter investigates the conditions of structural existence.

The two perspectives are therefore complementary rather than competing.

VII. Toward a Unified Structural Evolution Equation

To provide a unified description of

- generation;

- closure;
- feedback;
- and boundary,

the following section introduces the **Structural Evolution Master Equation.**

This equation does not belong exclusively to mathematics, physics, or chemistry. Instead, it serves as a cross-disciplinary framework for describing structural evolution.

Its purpose is not to replace existing scientific theories but to establish a common structural language shared among different scientific disciplines.

VIII. Concluding Remarks

Scientific progress continually demands new mathematical languages. As scientific inquiry shifts from **quantity** to **structure**, and from **dynamics** to **existence**, new mathematical representations naturally become necessary.

Concluding Statement:

Traditional mathematics describes how the world changes. Structural mathematics seeks to explain why stable structures are able to emerge and persist. The Structural Periodic Table therefore requires not merely new equations, but a new mathematical language for describing structural existence itself.

Section 3. A Unified Structural Evolution Master Equation — A Mathematical Formulation for Hierarchical Structural Stability

Abstract:

This section proposes a mathematical framework for hierarchical structural stability, providing a unified adjudication mechanism for the **Structural Periodic Table** and multi-level structural evolution.

Starting from the unified Structural Evolution Master Equation

$$\frac{\{d\Phi_L\}}{\{dt\}} = \alpha_1 \nabla \Phi_L + \alpha_2 \mathbf{I}(\Phi_L) + \alpha_3 \mathbf{Q}_L(\mathbf{t})$$

we introduce three complementary structural closure criteria:

- **Local Closure**, representing the controllability of irreducible structural units (Point •);
- **Dynamic Closure**, describing directional locking during structural evolution (Line |);
- **Global Closure**, expressing multi-scale structural consistency through feedback (Circle O).

These three criteria are unified into a Structural Lyapunov Functional,

$$\{\mathrm{L}\}_L \coloneqq a\{D\}_L + b\{P\}_L + c\Delta_L + r\{R\}_L$$

which serves as the mathematical adjudication function for structural stability.

We show that a hierarchical structure is regarded as stable whenever this functional remains **piecewise non-increasing** during both continuous evolution and discrete reconstruction events.

Furthermore, introducing the net structural dissipation rate

$$\kappa_L = \kappa_{\{L\}}^{\{\cdot\}} + \kappa_{\{L\}}^{\{1\}} + \kappa_{\{L\}}^{\{0\}} - \{\mathbf{NoiseGrowth}\}(\mathbf{L})$$

allows the maximal stable hierarchy to be defined as

$$L_{\{max\}} = \sup\{ L : \kappa_L > 0 \}$$

This formulation explains why higher-order hierarchical systems inevitably become unstable as structural complexity increases.

The proposed framework suggests that hierarchical limits are not empirical observations but mathematical consequences of the competition among **dissipation, branching complexity, and global feedback**. It therefore provides a general adjudication framework applicable to mathematics, physics, chemistry, and higher-order structural sciences.

I. Basic Objects and Notation

Let the structural state of a system at hierarchy level (L) be

$$\Phi_L(t) \in S_L$$

where (S_L) denotes the structural state space at hierarchy (L), whose degrees of freedom, constraints, and feedback channels all increase with hierarchical complexity.

The universal Structural Evolution Master Equation is proposed as

$$\frac{\{d\Phi_L\}}{\{dt\}} = \alpha_1 \nabla\Phi_L + \alpha_2\, I(\Phi_L) + \alpha_3 Q_L(t)$$

The three terms have the following structural meanings:

- $\nabla\Phi_L$: local perturbation or structural roughness (Point •);
- $I(\Phi_L)$: closure, gluing, surgery, or structural selection operator (Circle O);
- $Q_L(t)$: normalization, external driving, and scale continuation (Line |).

II. Three Structural Closure Criteria

2.1 Local Closure (Point •)

Define a structural roughness measure

$$D_L(\Phi) = |\nabla\Phi|^2$$

Local structural closure requires the existence of a positive constant (c_1) such that

$$\frac{\{d\}}{\{dt\}} D_L(\Phi_L(t)) \leq -c_1 D_L(\Phi_L(t)) + \eta_L(t)$$

where $\eta_L(t)$ represents unavoidable environmental noise.

If the effective noise grows faster than local dissipation as hierarchy increases, local closure fails.

2.2 Dynamic Closure (Line |)

Define structural path complexity

$$P_L(\Phi) = \log N_{paths}(\Phi; L)$$

or equivalently an effective branching rate $B_L(t)$

Dynamic closure requires

$$\sup_{\{t\geq 0\}} B_L(t) \leq B^*$$

or, more generally,

$$\frac{\{d\}}{\{dt\}} P_L \leq -c_2 P_L + \xi_L(t)$$

As hierarchical complexity increases, branching pathways multiply rapidly.

Without sufficient directional locking, structural evolution becomes non-repeatable and therefore non-adjudicable.

2.3 Global Closure (Circle O)

Define a multi-scale consistency error

$$\Delta_L(\Phi) := \sum_{k=1}^{K(L)} w_k \, d\left(\Pi_{\{k\}}(\Phi), \Pi_{\{0\}}(\Phi)\right)^2$$

Where

- $\Pi_{\{k\}}$ denotes structural projections from different models or scales,
- $d(.,.)$ is the structural distance,
- w_k are weighting coefficients.

Global closure requires

$$\frac{\{d\}}{\{dt\}} \Delta_L \le -c_3 \Delta_L + \xi_L(t)$$

together with the reconstruction condition

$$\Delta_L(t_j^+) - \Delta_L(t_j^-) \le 0$$

Failure of multi-scale consistency implies the breakdown of global structural closure.

III. Structural Lyapunov Functional

The three closure conditions are unified into a single structural energy functional,

$$\{L\}_L = a\{D\}_L + b\{P\}_L + c\Delta_L + r\{R\}_L$$

where $\{R\}_L$ represents normalization constraints such as scale conservation, boundary consistency, and global regularization.

Continuous stability requires

$$\frac{\{d\}}{\{dt\}}\{L\}_L \leq -\kappa\{L\}_L$$

Discrete reconstruction requires

$$\Delta\{L\}_L \leq 0$$

Together these conditions constitute the mathematical expression of the **Fourth Structural Cycle (Structural Adjudication)**:

$(\{L\}_L)\{$ piecewise non-increasing $\}\Longrightarrow\{$Stable Structural Existence$\}$.

IV. Emergence of the Hierarchical Stability Limit

Define the stable solution set

$\{Stable\}(L) = \{\ \Phi_L(0) : \{L\}_L(t)$ remains bounded and piecewise non-increasing$\}$

The maximal stable hierarchy may be characterized in three equivalent ways.

1) Collapse of the Stable Solution Space

$$\mu\big(\{\mathbf{Stable}\}(\mathbf{L})\big) \rightarrow \mathbf{0}$$

where μ denotes an appropriate measure of admissible initial conditions.

2) Loss of the Minimum Structural Energy

$$\inf_{\{\Phi\in\{S\}_L\}} \{L\}_L(\Phi)$$

ceases to exist or becomes unbounded beyond

$$L_{\{max\}}$$

3) Sign Reversal of the Net Dissipation Rate

Define

$$\kappa_L = \kappa_{\{L\}}^{\{\cdot\}} + \kappa_{\{L\}}^{\{1\}} + \kappa_{\{L\}}^{\{0\}} - \{\mathbf{NoiseGrowth}\}(\mathbf{L})$$

Then

$$\kappa_L > 0 \Rightarrow \{\text{Stable}\};$$
$$\kappa_L \leq 0 \Rightarrow \{\text{Unstable}\}$$

Consequently,

$$L_{\{max\}} = \sup\{ L : \kappa_L > 0 \}$$

This provides a mathematical explanation for the emergence of hierarchical instability at sufficiently high complexity.

As hierarchy increases, structural noise, branching complexity, and feedback interference eventually exceed the net dissipative capacity, causing κ_L to change sign.

V. Concluding Statement

Hierarchical stability is not determined by how many additional levels can be constructed, but by whether a piecewise non-increasing Structural Lyapunov Functional exists. Once the net structural dissipation rate satisfies ($\kappa_L \leq 0$), the hierarchy is structurally adjudicated as non-existent.

The Structural Evolution Master Equation describes how structures evolve; the Structural Lyapunov Functional determines whether they are permitted to exist. Together they constitute the mathematical adjudication framework of the Structural Periodic Table.

Section 4. First Structural Criterion (Generativity)

The Generativity Criterion for Chemical Reaction Networks — From Path Number to Generative Closure

Abstract:

As a structural example from chemistry, the first three cycles of a chemical reaction system show how the "possibility" of reaction is progressively closed through local interaction (Point •), pathway selection (Line │), and reproducible evolution (Circle O).

However, when the reaction system enters the fourth structural cycle, a higher-order question arises:

Does a reaction network possess genuine generativity?

In other words, can it continuously produce new structural states, rather than merely cycling within a finite set of channels?

To address this question, this section introduces two computable indicators:

$$\mathbf{P_L = \log N_{paths}}$$

as the **path-generativity index**, and $\{\mathbf{L}\}_\mathbf{L}$ as a Lyapunov-type **structural stability index**.

Together, these two quantities establish a joint adjudication framework for determining whether a chemical reaction network possesses true structural generativity.

I. Basic Definitions

1.1 Definition 1: Reaction Network

A chemical reaction network is represented as

$$\{G\} = (V,E)$$

where:

• (V) denotes the set of chemical species, including molecules and intermediates;

• (E) denotes the set of directed edges representing possible reactions;

• each edge (**e ∈ E**) is assigned a weight ($\mathbf{w_e}$), representing experimental support, a rate proxy, or a feasibility score.

1.2 Definition 2: Feasible Path

Given:

- a starting set ($S \subset V$),
- a target set ($T \subset V$),
- a maximum path length ($\mathbf{k_{\{max\}}}$),
- and a feasibility threshold ($\boldsymbol{\theta}$),

define

$$\mathbf{N_{paths}\ (S \rightarrow T;\ k_{\{max\}}, \theta\ )}$$

as the number of directed paths satisfying:

- path length ($\leq \mathbf{k_{\{max\}}}$);
- every edge weight ($\mathbf{w_e} \geq \boldsymbol{\theta}$).

1.3 Definition 3: Path-Generativity Index

Define

$$\mathbf{P_L = \log N_{paths}}$$

The logarithm is used in order to:

- avoid exponential blow-up;

• interpret the quantity as a hierarchical growth rate rather than a raw path count.

II. Stability Index: A Lyapunov-Type Quantity

Generativity without stability constraints may lead to structural collapse or random explosive growth.

Therefore, define a network stability index as

$$\{\mathbf{L}\}_{\mathrm{L}} = -\frac{1}{|V|}\sum_{\{i\in V\}}|\{\mathbf{dc_i}\}/\{\mathbf{dt}\}|$$

or, more generally,

$$\{\mathbf{L}\}_{\mathrm{L}} = -\nabla \cdot \mathbf{F}(\mathbf{c})$$

where:

- $\mathbf{c_i(t)}$ denotes the concentration of species (i);
- $\mathrm{F(c)}$ denotes the dynamical flow of the reaction network.

Interpretation:

- if ($\{\mathbf{L}\}_{\mathrm{L}} < 0$), the system converges toward an attractor and is structurally stable;
- if ($\{\mathbf{L}\}_{\mathrm{L}} > 0$), the system diverges and is structurally unstable.

III. Core Proposition: Generativity Adjudication

Proposition 1: Criterion for Generative Existence

A reaction network enters the **fourth-cycle generative level** if and only if

$$\mathbf{P_L} > \boldsymbol{\tau_1} \quad \text{and} \quad \{\mathbf{L}\}_{\mathrm{L}} < 0$$

where ($\boldsymbol{\tau_1}$) denotes the minimum threshold for path complexity.

This condition means that:

- the number of feasible pathways is sufficiently large, giving the network generative potential;
- the overall dynamical flow remains convergent, preventing structural collapse.

If only the first condition is satisfied, the network becomes an explosive random network.

If only the second condition is satisfied, the network remains a closed cyclic network.

True structural generativity requires both conditions to hold simultaneously.

IV. Algorithmic Procedure

4.1 Step 1: Construct the Network

- Construct (V) and (E) from experimental data.
- Estimate edge weights (w_e) using observed frequency, reaction rate, or feasibility score.

4.2 Step 2: Pruning

Choose a threshold (θ) and retain only edges satisfying

$$E_\theta = \{e \in E: \quad w_e \geq \theta\}$$

4.3 Step 3: Path Counting

Let (A) be the adjacency matrix of the pruned network.
Then

$$N_{\{paths\}} = \sum_{\{k=1\}}^{\{k_{\{max\}}\}} \sum_{\{s\in S\}} \sum_{\{t\in T\}} (A^k)_{\{s,t\}}$$

Therefore,

$$P_L = \log N_{paths}$$

4.4 Step 4: Stability Estimation

- Fit (dc/dt) from concentration time-series data.
- Estimate the divergence of the dynamical flow or the average convergence rate.
- Obtain ($\{L\}_L$).

4.5 Step 5: Joint Adjudication

- If (P_L) is small, the network is non-generative.
- If ($\{L\}_L > 0$), the network undergoes unstable explosive growth.
- If both criteria are satisfied, the network enters the fourth-cycle generative level.

V. Technical Notes

Several technical restrictions are necessary.

- Path number must be limited by ($k_{\{max\}}$); otherwise it may diverge.
- The edge-weight threshold must be based on experimental detection limits.
- Extremely weak pathways should not be counted as genuine generative channels, otherwise (P_L) may be artificially inflated.
- Stability estimation must be based on real concentration-evolution data.

VI. A Simple Example

Consider the reaction network

$A \to B,\ A \to C,\ B \to D,\ C \to D$

Let $(S = \{A\})$, $(T = \{D\})$, $(k_{\{max\}} = 2)$

There are two feasible paths:

1. $(A \to B \to D)$
2. $(A \to C \to D)$

Thus, $N_{paths} = 2$ and $P_L = \log 2$

If concentration evolution shows convergence toward (D), then $\{L\}_L < 0$ Therefore, the network satisfies the generativity criterion.

VII. Connection with the Unified Structural Evolution Equation

In the unified structural evolution equation,

$$\frac{\{dM\}}{\{dt\}} = \alpha_1 \nabla M + \alpha_2 I(E,S,C) + \alpha_3 Q(x)$$

we may identify

$$\alpha_3 Q(x) \leftrightarrow \{P\}_L$$

That is, the third term represents the generative source term, whose strength is determined by the path complexity index $\{P\}_L$.

Similarly,

$$\alpha_1 \nabla M \leftrightarrow \{L\}_L$$

which represents the structural stability or convergence term.

Thus,

{Fourth-cycle generativity} = {Path Complexity} + {Stable Convergence}

VIII. Summary

The generative capacity of chemical space is not determined merely by the number of molecules.

Rather, it is determined by the joint adjudication of feasible path complexity and stable convergence.

Only when a reaction network possesses both sufficient pathway richness and stable dynamical convergence can it be regarded as entering the fourth-cycle level of structural generativity.

Section 5. Second Structural Criterion (Closure Consistency)

A Multiscale Consistency Criterion for Molecular Geometry — Unifying Local Bond Constraints and Global Geometric Tension

Abstract:

As a chemical example, the first three structural cycles of molecular systems establish that

- chemical bonding constitutes the first threshold of structural closure (Circle O);
- geometric stability constitutes the second threshold;
- structural existence constitutes the third threshold.

However, once a molecular system enters the fourth structural cycle, a deeper question arises: **Can a theoretically admissible molecular geometry remain consistent across multiple structural scales?** In reality,

- many valence-allowed geometries cannot be maintained spatially;
- many geometrically drawable structures are dynamically unstable;
- numerous combinatorially possible molecules are naturally prohibited from synthesis.

To characterize this phenomenon, we introduce $\boldsymbol{\Delta_L}$ as a quantitative measure of **multiscale structural inconsistency**, providing a unified criterion for structural closure.

I. Basic Definitions

1.1 Definition 1: Local Bond Constraints

Consider a molecule containing (N) atoms. For atom (i),

- let ($\mathbf{C}_i^{local}$) denote its local bonding constraints;
- let $\boldsymbol{\theta}_{\{ij\}}^{(0)}$ be the ideal bond angle;
- let $\mathbf{r}_{\{ij\}}^{\{(0)\}}$ be the ideal bond length.

Define the local structural deviation

$$\boldsymbol{\delta}_i = \Sigma_j \left| \left\{ \mathbf{r}_{\{ij\}} - \mathbf{r}_{\{ij\}}^{(0)} \right\} / \left\{ \mathbf{r}_{\{ij\}}^{(0)} \right\} \right| + \Sigma_j \left| \boldsymbol{\theta}_{\{ij\}} - \boldsymbol{\theta}_{\{ij\}}^{(0)} \right|$$

This quantity measures the degree to which local bonding departs from its preferred structural configuration.

1.2 Definition 2: Global Geometric Tension

Define the global geometric tension as

$$\mathbf{T}_{\{global\}} = \Sigma_{\{loops\}} \left| \Sigma_{\{angles\}} \boldsymbol{\theta} - \boldsymbol{\Theta}_{\{closure\}} \right| + \Sigma_{\{steric\ pairs\}} \frac{\{1\}}{\{\mathbf{d}_{\{kl\}}\}}$$

where

- the first term measures deviations from geometric closure;
- the second term measures steric repulsion between non-bonded atomic pairs.

Together these characterize the global mechanical frustration of the molecular structure.

1.3 Definition 3: Multiscale Consistency Measure

Define $\boldsymbol{\Delta}_L = \Sigma_{\{i=1\}}^{N} \boldsymbol{\delta}_i + \boldsymbol{\beta} \mathbf{T}_{\{global\}}$

where ($\boldsymbol{\beta}$) is a scale-weighting parameter balancing local and global structural effects.

II. Core Proposition

Proposition 1 (Criterion for Structural Existence)

A molecular geometry enters the **fourth-cycle existence level** if and only if $\Delta_L < \varepsilon$

where (ε) denotes the structural tolerance threshold.

Whenever $\Delta_L > \varepsilon$, the molecular configuration is structurally adjudicated as non-existent, regardless of whether local bonding rules alone appear to permit it.

III. Structural Interpretation

The physical interpretation of Δ_L is summarized below.

Quantity	**Structural Meaning**
Local deviation	Point (•): local consistency
Global tension	Circle (O): global closure
Combined measure	Multiscale structural consistency

Table 1-4-5-1: Interpretation of the multiscale consistency measure.

Two typical failure modes arise naturally.

- Correct local bonding but incompatible global geometry leads to tension accumulation.
- Globally plausible geometry with excessive local distortion leads to bond instability.

Only the simultaneous satisfaction of **Local Consistency + Global Closure**, constitutes a genuinely admissible molecular structure.

IV. Computational Procedure

The proposed criterion can be evaluated using the following workflow.

Step 1: Input the molecular geometry obtained experimentally or computationally.

Step 2: Compute each local deviation $\boldsymbol{\delta_i}$

Step 3: Evaluate the global geometric tension $\mathbf{T_{\{global\}}}$

Step 4: Calculate $\boldsymbol{\Delta_L}$

Step 5: Perform structural adjudication.

- If ($\Delta_L \approx 0$): the structure is highly stable.
- Moderate values indicate metastable structures.
- Large values imply structural prohibition.

V. Illustrative Examples

5.1 Example 1. Water (H_2O)

The experimentally observed bond angle is approximately 104.5°. Suppose one artificially forces the molecule into a linear geometry 180°, Then

- local deviations increase substantially;
- global geometric tension also increases.

Consequently, $\Delta_L \uparrow$ and the linear configuration is structurally rejected.

5.2 Example 2. Methane (CH_4)

The tetrahedral geometry minimizes global geometric tension. Any planar configuration inevitably increases steric frustration. Therefore,

$$\Delta_{L_{\{\{tetrahedral\}\}}} < \Delta_{L_{\{\{planar\}\}}}$$

explaining why tetrahedral geometry is structurally selected.

VI. Connection with the Generativity Criterion

Within the unified structural evolution equation

$$\frac{\{dM\}}{\{dt\}} = \alpha_1 \nabla M + \alpha_2 I(E, S, C) + \alpha_3 Q(x)$$

the multiscale consistency measure controls the structural consistency component of I(E,S,C). Whenever (Δ_L) exceeds its admissible threshold, the generative term becomes suppressed.

A simple phenomenological relation is

$$Q(x) \propto e^{\{-\Delta_L\}}$$

Hence, the greater the structural consistency, the higher the probability of successful structural generation.

VII. Summary

Molecular geometry is not determined solely by valence-bond formulas. Rather, it is ultimately adjudicated by **multiscale structural consistency**.

Only structures satisfying both local bonding constraints and global geometric closure are capable of persistent existence.

Concluding Statement:

Chemical bonds determine what is locally possible; multiscale structural consistency determines what is globally allowed to exist. Within the Structural Periodic Table, molecular geometry is therefore understood not merely as a consequence of bonding rules, but as the outcome of structural adjudication across multiple hierarchical scales.

Section 6. Third Structural Criterion (Feedback Stability)

Periodicity as Structural Feedback Fixed Points — The Periodic Table as a Stable Feedback Attractor

Abstract:

In the chemistry volume of this work, we have established that

- elements represent the completion of the Point (•) structural role;
- elemental identity arises from localized electronic organization;
- elemental existence is constrained by multiscale structural consistency.

Yet a deeper question remains. Why do element families recur?
Why is periodicity stable rather than accidental?
Why is the number of stable elements finite rather than unlimited?
This section proposes a structural interpretation: **{Periodic Table} = {Map of Structural Feedback Fixed Points}**

From this viewpoint, periodicity is not fundamentally a consequence of electron-filling rules. Rather, it emerges as the repeated convergence of multiscale structural consistency toward discrete hierarchical feedback fixed points.

I. Basic Definitions

1.1 Definition 1 (Hierarchical Level)

Let **L** denote the hierarchical level of an elemental structure, representing the structural scale associated with localized electronic organization.

1.2 Definition 2 (Elemental Structural Consistency)

Define the structural consistency measure

$\boldsymbol{\Delta_L^{(E)}}$ = {Local Electronic Stability} + $\boldsymbol{\gamma}$ × {Global Nuclear–Electronic Tension}, where

- the first term characterizes local electronic localization,
- the second term quantifies the global coupling between nuclear and electronic structures.

Whenever

$$\boldsymbol{\Delta_L^{(E)}} < \varepsilon$$

the hierarchy admits a stable elemental configuration.

1.3 Definition 3 (Hierarchical Feedback Mapping)

Define the structural feedback map

$$\mathbf{F(L) = L + \alpha \cdot e^{\left\{-\Delta_L^{(E)}\right\}}}$$

If $\mathbf{F(L^*) = L^*}$, then $\mathbf{L^*}$ is called a **structural feedback fixed point**.

II. Core Structural Propositions

2.1 Proposition 1 (Periodic Fixed-Point Principle)

Element families recur if and only if $\boldsymbol{\Delta_L^{(E)}}$ reaches local minima at discrete hierarchical levels.

That is,

$$\{\mathbf{L_1, L_2, L_3, \ldots}\}$$

where each minimum corresponds to a structurally stable family within the periodic table.

2.2 Proposition 2 (Origin of Periodicity)

The periodic behavior of the elements is not fundamentally produced by electron-filling sequences. Instead,

{Periodicity} = {Repetition of Discrete Structural Feedback Fixed Points}.

III. Structural Interpretation

Within the fourth structural cycle,

Level	Structural Role
Point (•)	Local elemental stability
Line (|)	Hierarchical evolution
Circle (O)	Periodic structural closure
Fourth Cycle	Repetition of feedback fixed points

Table 1-4-6-1: Hierarchical interpretation of elemental periodicity.

Consequently, the periodic table may be viewed as a distribution of structural feedback fixed points within hierarchical structural space.

IV. Computational Framework

The proposed criterion may be implemented through the following procedure.

Step 1: Specify the hierarchical level L.
Step 2: Evaluate local electronic consistency, for example,

- electron-density uniformity,
- electronic repulsion consistency.

Step 3: Evaluate nuclear–electronic coupling, such as

- charge-density gradients,
- local field deviations.

Step 4: Compute $\Delta_L^{(E)}$
Step 5: Locate structural minima satisfying

$$\frac{\{d\}}{\{dL\}} \Delta_L^{(E)} = 0$$

These minima identify the structural positions of elemental families.

V. Illustrative Examples

5.1 Example 1. Noble Gases

When the electronic distribution becomes maximally uniform and structural tension reaches its minimum,

$$\Delta_L^{(E)} \rightarrow 0$$

a closed-shell configuration naturally emerges.
This corresponds to a structural closure point within the periodic table.

5.2 Example 2. Alkali Metals

At the beginning of each new hierarchical period,

- local structural stability is relatively weak,

- structural tension is comparatively high,
- chemical reactivity is consequently enhanced.

Nevertheless, their repeated occurrence at equivalent hierarchical positions generates the recurring alkali-metal family.

VI. Natural Emergence of the Elemental Upper Limit

As the hierarchical level continues to increase,

- nuclear–electronic tension grows,
- structural consistency becomes progressively more difficult to maintain.

Eventually, there exists $L_{\max}$ such that

$$\Delta_L^{(E)} > \varepsilon$$

Beyond this limit, stable elemental structures can no longer exist. Thus, the finite extent of the periodic table follows naturally from the breakdown of structural feedback stability.

VII. Connection with the Unified Structural Evolution Equation

Recall the unified structural evolution equation

$$\frac{\{dM\}}{\{dt\}} = \alpha_1 \nabla M + \alpha_2 I(E, S, C) + \alpha_3 Q(x)$$

At the elemental hierarchy,

- ∇M represents gradients in electronic organization;
- **I(E,S,C)** represents structural consistency;
- **Q(x)** represents hierarchical advancement.

Whenever $\Delta_L^{(E)}$ reaches a minimum, the evolution satisfies

$$\frac{\{dM\}}{\{dt\}} = 0$$

indicating convergence to a stable structural feedback fixed point.

VIII. Summary

The periodic table is not fundamentally an ordering of electron configurations. Rather, it is the spatial distribution of stable structural feedback fixed points generated through multiscale structural consistency.

Concluding Statement:

Electron filling describes the mechanism of periodicity; structural feedback explains its origin. From the perspective of the Structural Periodic Table, periodic recurrence emerges because multiscale structural systems repeatedly converge toward discrete feedback fixed points within hierarchical structural space.

Section 7. Fourth Structural Criterion (Finite Boundary)

Why Periodicity Possesses a Stable Upper Limit — A Finite-Boundary Theorem for Structural Feedback Fixed Points

Abstract:

The previous section established that periodic recurrence arises from repeated structural feedback fixed points.

Specifically, we introduced

- the hierarchical level (L),
- the structural consistency deviation ($\Delta_L^{(E)}$),
- the admissibility criterion

($\Delta_L^{(E)}\} < \varepsilon$)

and showed that elemental families emerge whenever $\Delta_L^{(E)}$ attains sufficiently small local minima (equivalently, whenever the feedback mapping admits a fixed point).

Consequently, the number of repetitions of elemental families is fundamentally determined by the number of sufficiently deep minima of $\Delta_L^{(E)}$ along the hierarchical axis.

This section proves that such minima can occur only finitely many times.

I. Fundamental Structural Decomposition

For every hierarchical structural system,
the structural consistency deviation admits the decomposition

$$\Delta_L^{(E)} = A(L) + B(L)$$

where

• **A(L)** denotes the **local organizational cost** (Point •), representing the structural cost required to maintain localized electronic stability;

• **B(L)** denotes the **global coupling cost** (Circle O), representing the energetic cost of maintaining nuclear–electronic consistency across increasing hierarchical scales.

1.1 Local Organizational Cost

Local structural organization may be improved through local rearrangement and structural closure.

However, its reduction possesses a lower bound,

$$\mathbf{A(L) \geq 0}$$

with

$$\mathbf{\inf A(L) = A_{\{min\}} \geq 0}$$

Thus, local optimization alone can never eliminate structural cost completely.

1.2 Global Coupling Cost

As hierarchical level increases,

• interaction range expands,
• multiscale coupling becomes stronger,
• maintaining global consistency becomes progressively more expensive.

Consequently, there exists some $\mathbf{L_0}$, such that for every $\mathbf{(L > L_0)}$,

$$B(L+1) - B(L) \geq \eta > 0$$

In other words, global structural tension grows monotonically beyond sufficiently high hierarchical levels.

This represents the weakest assumption required for irreversible structural accumulation.

II. Existence of a Finite Structural Boundary

Since $A(L) \geq \mathbf{A}_{\{min\}}$ while $B(L) \to \infty$, there necessarily exists a finite level $(\boldsymbol{L}_{\mathbf{max}})$, satisfying

$$\mathbf{A}_{\{min\}} + \mathbf{B}\left(\mathbf{L}_{\{max\}}\right) \geq \boldsymbol{\varepsilon}$$

Hence, for every $L \geq \mathbf{L}_{\{max\}}$,

$$\boldsymbol{\Delta}_{\mathbf{L}}^{(\mathbf{E})} = \mathbf{A(L) + B(L)} \geq \mathbf{A}_{\{min\}} + \mathbf{B(L)} \geq \boldsymbol{\varepsilon}$$

Therefore,

$L \geq \mathbf{L}_{\{max\}} \to$ {Elemental structures become structurally forbidden.}

This establishes that the admissible hierarchy of stable elements is necessarily finite.

III. Finiteness of Periodic Families

Element families correspond to local minima of $(\boldsymbol{\Delta}_{\mathbf{L}}^{(\mathbf{E})})$ namely,

$$\boldsymbol{\Delta}_{\{\mathbf{L_k}\}}^{\{(\mathbf{E})\}} < \boldsymbol{\varepsilon}, \quad \text{together with} \quad \boldsymbol{\Delta}_{\{\mathbf{L_k}\}}^{\{(\mathbf{E})\}} \leq \boldsymbol{\Delta}_{\{\mathbf{L_k \pm 1}\}}^{\{(\mathbf{E})\}}$$

However, the previous theorem demonstrates that

$$\boldsymbol{\Delta}_{\mathbf{L}}^{(\mathbf{E})} < \boldsymbol{\varepsilon}$$

can occur only within the finite interval

$$\mathbf{L} \in \left\{\mathbf{1, 2, \ldots, L_{\{max\}} - 1}\right\}$$

A finite interval contains only finitely many local minima.

Therefore, {**The repetition of elemental families must itself be finite.**}

IV. Finite Number of Structural Feedback Fixed Points

Recall the structural feedback mapping

$$\mathbf{F(L) = L + \alpha e^{\{-\Delta_L^{\{(E)\}}\}}}$$

A feedback fixed point requires sufficiently small structural deviation so that the feedback term can compensate for hierarchical drift.

However, once $(L \geq \mathbf{L_{\{max\}}})$ one has $\mathbf{\Delta_L^{(E)} \geq \varepsilon}$ and therefore

$$\mathbf{e^{\{-\Delta_L^{\{(E)\}}\}} \leq e^{\{-\varepsilon\}}}$$

The feedback amplitude then enters a weak-feedback regime, making further structural locking impossible.

Consequently, the number of structural feedback fixed points is necessarily finite,

$$\{|\{\mathbf{L: F(L) = L}\}| < \infty\}$$

V. Finite-Boundary Theorem

The finite repetition of elemental families is not merely an empirical feature of the periodic table.

Rather, it follows inevitably from the structural feedback mechanism.

As hierarchical level increases, global structural tension grows irreversibly, forcing the multiscale consistency deviation $(\mathbf{\Delta_L^{(E)}})$ to exceed its admissible threshold.

Beyond this point, no further feedback fixed points can exist, and periodic recurrence necessarily terminates.

VI. Summary

The periodic table possesses a finite extent because structural feedback admits only a finite number of stable fixed points.

Periodicity is therefore not an indefinitely repeatable filling process, but the finite recurrence of structurally admissible feedback attractors.

Concluding Statement:

The repetition of elemental families is not the consequence of unlimited electronic filling. It is the finite recurrence of structural feedback fixed points. As hierarchical complexity increases, accumulated structural tension eventually interrupts global closure, thereby establishing the finite boundary of periodicity.

Finite-Boundary Theorem:

The Structural Periodic Table admits only finitely many structurally stable feedback fixed points. Therefore, every periodic hierarchy necessarily possesses a finite structural boundary.

Section 8. Principle of Structural Adjudication: From Verification to Prediction

Abstract:

The previous chapter established the scientific verification framework for the Structural Periodic Table. This chapter has further developed a unified mathematical language of **structural adjudication**, introducing four fundamental criteria:

- Generativity,
- Structural Closure,
- Feedback Stability,
- Finite Boundary.

These four criteria are not independent mathematical conditions. Rather, together they describe the complete evolutionary process through which structures emerge, stabilize, and undergo hierarchical transitions.

Structural adjudication is therefore more than a mathematical description. It provides a unified cross-disciplinary principle for determining the conditions under which structures can persist, evolve, and enter higher levels of organization.

I. A Unified Framework for the Four Structural Criteria

The four structural criteria proposed in this chapter correspond to four successive stages of structural evolution.

1) First Criterion: Generativity

Question: Can a new structure emerge?
Primary focus: Mechanisms of structural generation.

2) Second Criterion: Structural Closure

Question: Can the generated structure organize into a coherent and stable whole?

Primary focus: Structural closure and multiscale consistency.

3) Third Criterion: Feedback Stability

Question: Can a stable structure continuously maintain itself?

Primary focus: Feedback mechanisms and structural fixed points.

4) Fourth Criterion: Finite Boundary

Question: Why does structural growth eventually terminate, giving rise to a new hierarchical level?

Primary focus: Existence boundaries and structural limits.

Consequently, structural evolution is not an unbounded continuous process. Instead, new hierarchical levels emerge only after the completion of a finite structural cycle.

II. From Mathematical Description to Structural Adjudication

The unified Structural Evolution Master Equation introduced in this chapter is not intended to replace existing mathematical theories.

Its purpose is to provide a common language for describing structural existence across different scientific disciplines.

Traditional mathematics primarily asks: **How does a system evolve?**

Structural adjudication further asks: **Why can a structure exist?**

Accordingly, structural mathematics investigates not only dynamical evolution, but also the conditions that permit stable existence.

III. From Scientific Verification to Theoretical Prediction

Chapter Three established the framework for scientific verification.

Chapter Four establishes the mathematical framework for structural adjudication.

Together, they constitute the theoretical foundation of the Structural Periodic Table. Whenever a structure simultaneously satisfies

- Generativity,
- Structural Closure,
- Feedback Stability,
- Finite Boundary,

it is not only experimentally admissible, but also capable of entering a higher structural hierarchy.

Structural adjudication therefore extends beyond explaining existing structures. It provides a framework for predicting structures that have not yet emerged.

IV. Toward Verification in the Natural Sciences

The first four chapters establish the theoretical foundations of the Structural Periodic Table. The following chapters apply these principles separately to mathematics, physics, and chemistry.

- Mathematics examines the logical consistency of structural language.
- Physics investigates the correspondence between structural principles and natural laws.
- Chemistry explores the realization of structural principles within material systems.

The substantive validation of the Structural Periodic Table therefore begins in the chapters that follow.

V. Summary

Structural adjudication is not a simple evaluation of individual theories. Rather, it determines whether a structure possesses the capacity for persistent existence, stable development, and hierarchical advancement.

It does not primarily ask: **Is this theory correct?**

Instead, it asks: **Does this structure possess the conditions necessary for existence?**

Structural adjudication therefore serves as the essential bridge connecting theoretical construction, experimental verification, and scientific prediction.

Concluding Statements: Only after simultaneously satisfying the four structural criteria — **Generation, Structural Closure, Feedback Stability, and Finite Boundary**—can a structure complete a hierarchical transition and enter a new stable structural cycle.

Scientific verification determines whether a theory is empirically credible. Structural adjudication determines whether a structure is fundamentally admissible. Together, they establish the scientific foundation of the Structural Periodic Table.

A structure does not advance because it becomes more complicated; it advances because it successfully completes all four stages of structural adjudication.

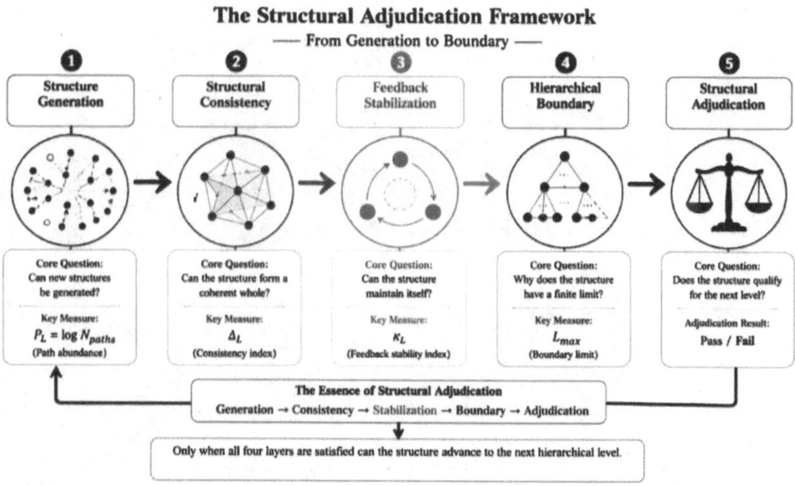

Figure 1-4-8-1: Structural Adjudication Framework

References: (Chapter Four)

(Selection of foundational, classic, and structurally stable papers)

[1] Lyapunov, A. M. (1992). *The General Problem of the Stability of Motion*. Taylor & Francis.

[2] Arnold, V. I. (1989). *Mathematical Methods of Classical Mechanics*. Springer.

[3] Smale, S. (1967). Differentiable Dynamical Systems. *Bulletin of the American Mathematical Society*, 73(6), 747–817.

[4] Thom, R. (1975). *Structural Stability and Morphogenesis*. W. A. Benjamin.

[5] Prigogine, I. (1980). *From Being to Becoming: Time and Complexity in the Physical Sciences*. Freeman.

[6] Haken, H. (1983). *Synergetics: An Introduction*. Springer.

[7] Strogatz, S. (2015). *Nonlinear Dynamics and Chaos*. Westview Press.

[8] Penrose, R. (2005). *The Road to Reality*. Vintage.

Part II — Theory and Validation of Structural Periodicity in the Natural Sciences

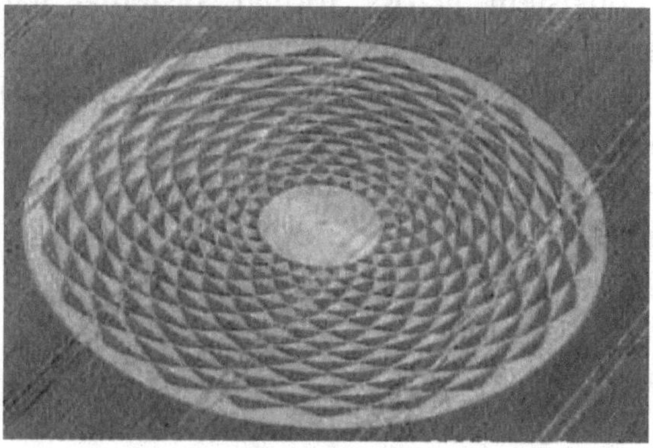

Figure 2: Crop Circle Illustrating a Galactic Civilization. See book 《*Crop Circle*》 for details.

Structural Logic of Part II

In this part, mathematics, physics, and chemistry are not treated merely as three separate disciplines. Rather, they play three distinct structural roles:

- **Mathematics** tests whether structure possesses **logical consistency**.
- **Physics** tests whether structure possesses **physical correspondence**.
- **Chemistry** tests whether structure possesses **material realization**.

Thus, these three disciplines are not simply examples. They constitute three different levels of evidence.

They form a verification chain moving from: **logic → nature → matter,** and ultimately toward life.

Accordingly, Part II develops the Structural Periodic Table through the sequence:

Mathematics → Physics → Chemistry → Life

This sequence examines whether structural periodicity can remain consistent across formal systems, natural laws, material organization, and the emergence of living systems.

In short, Mathematics provides the logical prototype; Physics provides the dynamical realization; Chemistry provides the material realization. Together they constitute the three foundational layers supporting the Structural Periodic Table.

Chapter Five: Theory of Mathematical Structural Periodicity

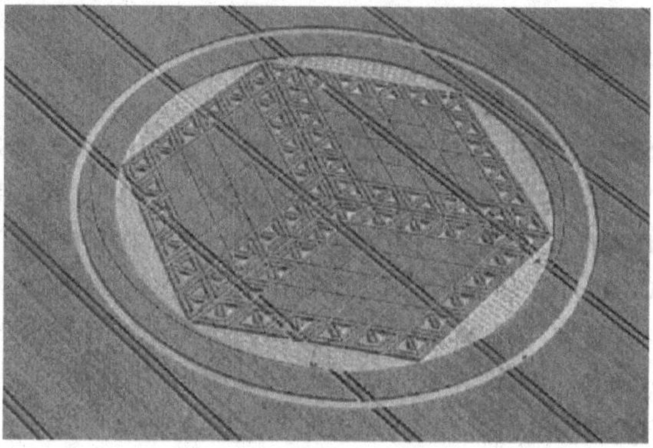

Figure 2-5: Crop Circle Illustrating a Galactic Civilization. See book 《*Crop Circle*》 for details.

Chapter Overview:

This chapter analyzes five fundamental mathematical structural nodes:

- **Recursive Structure** — corresponding to generation;
- **Groups and Symmetry** — corresponding to constraint;
- **Periodic Functions** — corresponding to repetition;
- **Moduli Spaces** — corresponding to classification;
- **Structural Spectra** — corresponding to stability.

Together, these five nodes form a mathematical hierarchy of structural generation. They may be summarized as the following mathematical structural chain:

Recursive Generation → Symmetric Constraint → Periodic Repetition → Moduli Classification → Spectral Stability

This chapter therefore treats mathematics not merely as a computational tool, but as the most abstract prototype of structural periodicity.

Section 1. Basic Mathematical Prototypes of Structural Periodicity

Abstract:

In structural periodic theory, mathematics is not merely a descriptive tool. It is the most stable and abstract layer of structure.

Compared with physics and chemistry, which depend on specific conditions of matter and energy, mathematical structures are closer to pure logical organization. For this reason, mathematics more clearly reveals fundamental structural features such as periodicity, closure, and stability windows.

If structural periodicity is understood as: **generation → constraint → repetition → classification → stability**

then mathematics provides a group of typical structural prototypes.

These prototypes are not isolated. They form five basic expressions of mathematical structural periodicity:

- recursive structures;
- groups and symmetry;
- periodic functions;
- moduli spaces;
- structural spectra.

Together, they provide the most fundamental language for cross-disciplinary structural periodicity.

I. Recursive Structure: The Basic Mechanism of Structural Generation

Recursion is one of the most fundamental generative mechanisms in mathematics. Through the repeated application of simple rules, complex structures can gradually emerge.

Typical examples include:

- the Fibonacci sequence;

- recursive functions and recursive definitions;
- fractal structures and self-similar patterns.

In structural language, recursion expresses the self-extension of local rules across multiple levels. Each recursive step generates a new level, while the generative principle remains unchanged.

Thus, recursion is not merely a computational tool. It is a form of structural generation. From the perspective of structural periodicity, recursion implies:

- continuous hierarchical expansion;
- repeated appearance of similar structures;
- preservation of local rules within global organization.

This mechanism corresponds deeply to self-organization in natural systems.

II. Groups and Symmetry: The Mathematical Expression of Structural Constraint

If recursion represents generation, then groups and symmetry represent constraint.In mathematics, symmetry is precisely expressed through group structures. Examples include:

- geometric symmetry groups;
- permutation groups;
- Lie groups and Lie algebras.

Group structures reveal which properties remain invariant under allowed transformations. These invariants form the foundation of structural stability.

From the perspective of structural periodicity, symmetry provides:

- conditions for structural closure;
- coordination between hierarchical levels;
- selection rules for possible structures.

Thus, group theory does not merely describe symmetry. It also describes which structures can exist stably.

III. Periodic Functions: The Mathematical Form of Repetition

Periodic functions are the most direct mathematical expression of periodicity. Typical examples include:

- trigonometric functions;
- Fourier expansions;
- periodic solutions of wave equations.

The defining property of a periodic function is

$$f(x+T) = f(x),$$

where (T) denotes the period.

This relation shows that structure can recur at a given scale.

In structural periodic theory, periodic functions not only describe repeated phenomena. They also reveal:

- stable rhythms of structure;
- oscillation and feedback mechanisms;
- superposition of multi-scale periodicities.

Periodic functions therefore provide an important tool for understanding rhythms in complex systems.

IV. Moduli Spaces: A Geometric Framework for Structural Classification

As the number of structures increases, recursion and symmetry alone are no longer sufficient to describe all possible forms. A higher-level framework becomes necessary for classifying structures. This is the role of moduli spaces.

The core idea of a moduli space is to treat objects with the same structural properties as belonging to the same class.

Examples include:

- moduli spaces of curves;
- moduli spaces of vector bundles;
- moduli spaces of complex structures.

Moduli spaces provide a geometric way to understand structural diversity:

- different structures are mapped into the same parameter space;
- structural change corresponds to paths within that space;
- stable structures correspond to special regions.

Thus, a moduli space may be understood as a distribution map of structural periodicity in higher-dimensional space.

V. Structural Spectra: The Final Expression of Stable Structure

In many mathematical and physical systems, structure ultimately appears as a set of discrete values. This phenomenon is commonly described through spectra.

Typical examples include:

- eigenvalue spectra of matrices;
- spectra of Laplace operators;
- energy spectra of quantum systems.

Spectral structures reveal that among many possible states, only certain values can remain stable.

From the perspective of structural periodicity, spectra represent:

- stable nodes of periodic systems;
- fixed points of feedback structures;
- discrete states after hierarchical closure.

Thus, structural spectra may be understood as the stable expression of completed structural periodicity.

VI. Summary

Through the five prototypes of recursive structure, groups and symmetry, periodic functions, moduli spaces, and structural spectra, we observe a basic evolutionary pathway of mathematical structural periodicity:

Recursive Generation → Symmetric Constraint → Periodic Repetition → Moduli Classification → Spectral Stability

This process demonstrates the full development of structural periodicity within mathematics.

Mathematics therefore provides not only computational tools, but also a fundamental understanding of how structures are generated, repeated, and stabilized.

In later chapters, we will see how similar periodic structures appear in physics and chemistry, ultimately contributing to the interdisciplinary Structural Periodic Table.

Section 2. A Triadic Interpretation of Mathematical Structural Periodicity

Abstract:

In the previous section, we examined five typical prototypes of mathematical structural periodicity: recursive structures, groups and symmetry, periodic functions, moduli spaces, and structural spectra.

Although these structures differ in form, from a deeper structural perspective they all express the interaction of three fundamental structural dynamics.

In the structural language of this book, these three dynamics are abstracted as:

- **Point •** — local generation and source of imbalance;
- **Line |** — evolutionary pathway and structural unfolding;
- **Circle O** — closure constraint and stable structure.

These three structural elements are not simple geometric metaphors. They are abstract expressions of three fundamental mechanisms: structural generation, propagation, and stabilization.

Within mathematics, many apparently complex structures can be interpreted through the interaction of these three structural dynamics.

I. Point •: Local Generation and Structural Origin

In mathematical structures, the Point corresponds to a local generative rule or fundamental element. Examples include:

- initial conditions of recursive functions;
- initial terms of a sequence;
- identity elements in group structures;
- basic point sets in geometry.

These objects provide the starting point for structural generation. Without such local origins, no hierarchical structure can unfold.

In recursive systems, a simple initial rule can generate highly complex structures. For example, the Fibonacci sequence satisfies

$$F_{\{n+1\}} = F_n + F_{\{n-1\}}$$

Its complex growth pattern arises from a very simple generative relation. From the perspective of structural periodicity, the Point represents local rules and the source of structural generation.

II. Line │: Structural Unfolding and Evolutionary Pathway

If the Point provides the starting point of generation, the Line represents the pathway of structural unfolding.

In mathematics, this pathway may appear as:

- stepwise generation of recursive sequences;
- evolution of functions in parameter space;
- orbits under group actions;
- trajectories of solutions to differential equations.

These structures express the evolution of mathematical objects under rule-governed processes. For example, in group theory, the orbit of an element under a group action forms a complete structural pathway. Similarly, in dynamical systems, solution trajectories describe how a system evolves over time.

From the perspective of structural periodicity, the Line represents the unfolding and propagation of structure under generative rules.

III. Circle O: Closed Structure and Stable State

After generation and evolution, a structure becomes stable only if it can form closure. In mathematics, such closure may appear as:

- closed operations in group structures;
- repeated periods in periodic functions;
- stable equivalence classes in moduli spaces;

- discrete stable values in operator spectra.

For example, a periodic function satisfies

$$f(x+T) = f(x).$$

This closure relation indicates that structure repeats at a given scale. Likewise, in spectral theory, stable structures often correspond to discrete eigenvalues. These form stable modes of a system.

Thus, the Circle represents stable closure formed under structural constraints.

IV. Coupling Among the Three Structural Roles

Point, Line, and Circle are not independent structures. They form a cyclic system: **Point → Line → Circle → New Point**

This cycle can be understood as follows:

- local rules generate structure (Point);
- structure unfolds along pathways (Line);
- structure forms closure under constraints (Circle);
- closed structure becomes the starting point for a new level (New Point).

This process constitutes the basic mechanism through which structural periodicity emerges.

In mathematics, many apparently different theoretical structures can be located within this cycle. For example:

- recursion generates new structural levels;
- symmetry provides closure conditions;
- periodic functions form repeating structures;
- moduli spaces classify structures;
- spectral structures provide stable solutions.

These phenomena are not isolated from one another. They are different expressions of the same triadic structural cycle across different areas of mathematics.

V. The Triadic Structure as a Cross-Disciplinary Structural Language

From a broader perspective, Point, Line, and Circle are not merely abstractions within mathematics. They also provide a unified language for understanding structural periodicity in other disciplines.

In later chapters, we will see that:

- in physics, local field sources, propagation modes, and geometric closure form dynamical structures;
- in chemistry, elements, reaction pathways, and molecular structures form material hierarchies;
- in life systems, individual generation, evolutionary processes, and ecological stability form cyclic organization.

Thus, the triadic structure is not a special concept belonging to one discipline. It is a basic model for describing multi-level structural generation and stability.

VI. Summary

Many structural phenomena in mathematics can be interpreted through the triadic framework of Point, Line, and Circle:

- **Point:** source of structural generation;
- **Line:** pathway of structural evolution;
- **Circle:** stable structural closure.

Together they form the cycle: **Generation → Evolution → Closure → New Hierarchical Generation**

This cycle constitutes the basic mechanism of structural periodicity.

In the following chapters, we will see how this triadic structure continues to unfold in physical and chemical systems, ultimately forming the interdisciplinary Structural Periodic Table.

Section 3. Summary: From Mathematical Structural Periodicity to Structural Periodicity in the Natural Sciences

This chapter began with mathematical structure and examined structural periodicity at its most abstract level.

Through the five prototypes of recursive structures, groups and symmetry, periodic functions, moduli spaces, and structural spectra, we identified the generative pathway of mathematical structural periodicity:

Recursive Generation → Symmetric Constraint → Periodic Repetition → Moduli Classification → Spectral Stability

This process shows that periodicity is not merely repetition. It is the result of structural circulation among generation, constraint, and stabilization.

Further analysis unified these mathematical structures into three fundamental structural dynamics:

- **Point •:** local generation and structural origin;
- **Line |:** structural unfolding and evolutionary pathway;
- **Circle O:** closure constraint and stable structure.

Together, they form the cycle: **Point → Line → Circle → New Point**

This cycle reveals the basic mechanism of structural periodicity:

- local rules generate a structural origin;
- structure unfolds along pathways and forms hierarchies;
- stable closure emerges under constraints;
- closed structures become the foundation for a new level of generation.

Recursive structures, symmetry, periodic functions, moduli spaces, and spectra in mathematics are all expressions of this triadic structural cycle in different domains.

Accordingly, this chapter reaches an important conclusion: **Mathematics is not only the language used to describe nature; it**

also reveals the basic principles by which structures are generated and stabilized.

This triadic structural language provides a unified framework for the following chapters.

- In physics, similar structures appear as relationships among local field sources, propagation modes, and geometric closure.
- In chemistry, they appear as hierarchical organization among elements, reaction pathways, and molecular structures.

In other words, mathematical structural periodicity is not an isolated phenomenon. It is the abstract prototype of structural periodicity in the natural sciences.

Thus, this chapter completes the transition: **Mathematical Structural Periodicity → Triadic Structural Language → Structural Periodicity in the Natural Sciences**

This transition establishes the theoretical foundation for the following discussions of physical and chemical structural periodicity.

Note: Mathematics reveals the logical foundations of structural periodicity. Physics demonstrates its dynamical realization. Chemistry embodies its material manifestation. Together, they establish the three foundational pillars of the Structural Periodic Table.

References: (Chapter Five)

The following references provide mathematical background for recursion, symmetry, periodicity, moduli spaces, and spectral structures. They situate the discussion within established mathematical theory while supporting the structural interpretation proposed in this chapter.

[1] Bourbaki, N. (1989). *Elements of Mathematics: Theory of Sets*. Springer.

[2] Mac Lane, S. (1998). *Categories for the Working Mathematician* (2nd ed.). Springer.

[3] Fulton, W. & Harris, J. (1991). *Representation Theory: A First Course*. Springer.

[4] Conway, J. B. (1990). *A Course in Functional Analysis* (2nd ed.). Springer.

[5] Arnold, V. I. (1992). *Ordinary Differential Equations*. Springer.

[6] Serre, J. P. (1977). *Linear Representations of Finite Groups*. Springer.

[7] Griffiths, P. & Harris, J. (1994). *Principles of Algebraic Geometry*. Wiley.

[8] Strogatz, S. H. (2015). *Nonlinear Dynamics and Chaos* (2nd ed.). Westview Press.

[9] Weyl, H. (1952). *Symmetry*. Princeton University Press.

[10] Chang, J. (2026). *Recursive Self-Organizing Evolutionary Breakthrough — Unifying Natural, Social and Life Sciences*. Universal Publishing.

Chapter Six: Unified-Theory Validation of Mathematical Structural Periodicity

Figure 2-6: Crop Circle Illustrating a Galactic Civilization. See book《*Crop Circle*》for details.

Chapter Overview:

The relationship between structural periodicity and unified theory brings the following themes into a single conceptual framework:

- Section 1: high-dimensional boundaries;
- Section 2: the structural adjudication framework;
- Section 3: the minimal unified structure;
- Section 4: case studies;
- Section 5: the triadic structural pattern;
- Section 6: the structural mapping principle of unified theories.

Together, these sections converge toward one central idea:

The Structural Periodic Table does not replace mathematical theories. Rather, it provides a common structural coordinate system through which different theories may be compared, mapped, and adjudicated.

This statement is very close to the central positioning of the entire book.

It is also necessary to clarify why this chapter, while discussing the application of mathematical structural periodicity, repeatedly refers to theories usually associated with physics.

The reason is that theories such as superstring theory and M-theory have physical ambitions, yet their internal structure is profoundly mathematical.

They occupy an intermediate region between mathematics and physics.

Ordinary physical theory typically investigates such domains as:

- electromagnetism;
- gravitation;
- quantum theory.

These will be discussed in Chapters 7 and 8.

By contrast, superstring theory and M-theory involve mathematical structures such as:

- Calabi–Yau manifolds;
- high-dimensional topology;
- duality transformations;
- fiber bundles;
- supersymmetry groups.

These are closer to mathematical structural theory than to direct experimental physics. Therefore, this chapter primarily examines the structural side of mathematics.

Its central task is not to verify whether a physical theory is empirically correct, but to ask whether a proposed unified theory possesses **logical consistency** as a mathematical structure.

In this sense, mathematics serves as the first level of validation in the Structural Periodic Table: **Mathematics tests whether a structure is logically coherent.**

Section 1. Structural Boundaries of High-Dimensional Unified Theories

From Dimensional Expansion to Structural Generation

Abstract:

The development of unified theories has long been driven by a central question: **how can a single framework explain an increasing range of physical phenomena?** One of the major directions in modern theoretical physics has been the introduction of higher-dimensional spaces, exemplified by the ten-dimensional framework of superstring theory and the eleven-dimensional framework of M-theory. Although these theories exhibit remarkable mathematical consistency, they also raise a more fundamental question:

Does the development of unified theory itself possess structural boundaries?

Rather than evaluating whether higher-dimensional theories are correct, this section addresses a more general issue: **Is the growth of theoretical complexity constrained by structural principles?**

From the perspective of the Structural Periodic Table, dimensionality is not the central issue. The essential question is whether additional structures play an **irreducible generative role**. If a theory must continually introduce new degrees of freedom merely to preserve internal consistency, then its claim to unification deserves renewed structural examination.

This section therefore proposes a structural-boundary perspective that lays the foundation for the adjudication framework developed in the following sections.

I. The Evolution of Unified Theories

Modern physics has exhibited a persistent tendency toward unification.

From classical mechanics to electromagnetic theory, from electroweak unification to Grand Unified Theories, physicists have continually sought increasingly comprehensive descriptions of nature.

Within this historical development, higher-dimensional theories have emerged as one of the most influential approaches.

Representative examples include:

- Superstring Theory (10 dimensions);
- M-Theory (11 dimensions);
- various compactified higher-dimensional models.

These theories attempt to achieve broader unification by introducing additional degrees of freedom.

From this perspective, higher-dimensional theories may be regarded as processes of structural expansion.

II. Does Increasing Dimensionality Necessarily Strengthen Unification?

History suggests that greater theoretical complexity does not necessarily imply greater unifying power.

For example:

- The Copernican system proved simpler than the Ptolemaic model.
- Maxwell's equations unified electricity and magnetism.
- Einstein unified gravitation through geometric structure.

These successful theories share a common feature: **They explained more phenomena with fewer fundamental structures.**

This naturally leads to an important question: **Are the additional dimensions introduced in higher-dimensional theories structurally indispensable?**

If extra dimensions merely preserve formal mathematical consistency without introducing fundamentally new generative mechanisms, then their role in unification should be reconsidered.

III. Structural Boundaries from the Perspective of the Structural Periodic Table

The Structural Periodic Table focuses not on the number of dimensions, but on mechanisms of structural generation.

From this viewpoint, every unified theory may encounter three fundamental structural boundaries.

3.1 Minimal Generative Boundary

Can the theory be reduced to fewer fundamental structural principles?

If further compression is possible, the present formulation may contain structural redundancy.

3.2 Stable Closure Boundary

Can the theory achieve stable structural closure?

If it produces enormous landscapes of equivalent solutions, infinitely many possible vacua, or an unlimited number of mathematically acceptable states, then the degree of structural closure requires further examination.

3.3 Predictive Boundary

Does the theory possess genuine generative predictive power?

A true unified theory should not merely explain known phenomena; it should also predict structures whose existence follows necessarily from its internal principles.

IV. The Structural Challenge Facing Higher-Dimensional Theories

From the structural perspective, higher-dimensional theories are not questioned simply because they involve many dimensions.

The central issue is whether those dimensions possess **irreducible structural significance**.

For example:

- Why ten dimensions rather than nine?
- Why eleven rather than twelve?
- Do the additional dimensions correspond to genuine mechanisms of structural generation?

If these questions cannot be answered, then higher-dimensional structures may represent mathematical expansion rather than structural generation.

Consequently, the Structural Periodic Table is concerned not with the quantity of dimensions, but with whether each dimension performs an indispensable structural role.

V. Structural Boundaries Do Not Reject Higher-Dimensional Theories

It should be emphasized that this discussion does **not** reject Superstring Theory, M-Theory, or any other higher-dimensional framework.

The Structural Periodic Table is not proposed as a replacement for existing theories. Instead, it offers an additional structural perspective from which theories may be examined.

If future developments demonstrate that higher-dimensional theories satisfy the following conditions:

- the additional dimensions possess irreducible generative functions;
- the theory achieves stable structural closure;
- the theory provides clear structural predictions;

then such theories may successfully pass structural evaluation.

Structural boundaries therefore do not restrict theoretical development. Rather, they help clarify the generative foundations upon which a unified theory is built.

VI. Conclusion

The evolution of unified theories does not necessarily require an unlimited increase in degrees of freedom.

From the perspective of the Structural Periodic Table, the more fundamental questions are whether a theory possesses:

- minimal generativity;
- stable structural closure;
- genuine predictive capability.

Accordingly, the essential question facing higher-dimensional theories is not: **"How many additional dimensions can be introduced?"**

but rather: **"Do these dimensions possess indispensable structural significance?"**

This question defines the structural boundary of higher-dimensional unified theories.

The next section will further examine whether the existence of such structural boundaries makes it possible to establish a structural adjudication framework that is independent of any particular physical theory.

Concluding Statement:

The boundary of a unified theory may not ultimately be determined by the number of dimensions it contains, but by its capacity for structural generation. Higher-dimensional organization alone does not guarantee deeper unification; genuine unification is achieved only when increasing complexity ultimately converges toward a coherent structural whole.

Section 2. A Theoretical Adjudication Framework for the Structural Periodic Table

From Higher-Dimensional Unified Theories to Principles of Structural Evaluation

Abstract:

This section proposes a theoretical adjudication framework based on the **Structural Periodic Table** for evaluating the structural legitimacy of higher-dimensional unified theories.

The Structural Periodic Table is not a classification of material elements, but a hierarchical framework of structural closure whose generative core is represented by the three fundamental structural roles: **Point (•), Line (|), and Circle (O).**

Three principles of structural adjudication are proposed:

- the Principle of Minimal Generation;
- the Principle of Unique Structural Closure;
- the Principle of Predictive Structural Vacancies.

These principles are applied conceptually to higher-dimensional unified theories, such as Superstring Theory, as well as to the author's proposed **Ultra-Metastructural Unification Theory (Point–Line–Circle Theory).**

A central premise of this chapter is that, if the Structural Periodic Table is valid, then every theory — including the author's own — must be evaluated according to exactly the same structural criteria.

Structure serves no particular theory; every theory must conform to structural principles.

I. Why Must Unified Theories Be Subject to Structural Evaluation?

Modern high-energy physics has produced numerous higher-dimensional approaches to unification, including:

- ten-dimensional Superstring Theory;
- eleven-dimensional M-Theory;
- compactified higher-dimensional spaces;
- multiple vacuum landscapes.

Although these theories possess remarkable mathematical sophistication, they lack a universally accepted structural criterion for evaluation.

The essential question is therefore not: **"How mathematically complex is the theory?"**

but rather: **"Does the theory achieve genuine structural closure?"**

If a theory can indefinitely expand its degrees of freedom without a corresponding mechanism of structural adjudication, it may represent an open mathematical framework rather than a genuinely generative structure.

For this reason, the Structural Periodic Table is proposed as a higher-level structural language for theoretical evaluation.

II. Three Principles of Structural Adjudication

The essence of the Structural Periodic Table lies not in dimensionality, but in structural generation.

2.1 Principle One: Minimal Structural Generation

Every legitimate theory should ultimately be reducible to the fundamental generative sequence: **Point (local irreducibility) → Line (evolution and coupling) → Circle (global closure).**

If a theory:

- requires continually introducing additional degrees of freedom in order to remain internally consistent; or
- cannot be reduced to this fundamental generative logic,

then structural redundancy may exist. Higher dimensionality is not itself problematic.

Irreducibility — not dimensionality — is the true structural criterion.

2.2 Principle Two: Unique Structural Closure

A structural framework should converge toward stable closure. If a theory produces:

- infinitely many landscapes;
- multiple equivalent vacua;
- no unique criterion for structural existence,

then it fails to satisfy the structural closure represented by the Circle (O). Complexity is acceptable. Arbitrariness is not.

2.3 Principle Three: Predictive Structural Vacancies

If the Structural Periodic Table is valid, it should possess predictive power analogous to that of the chemical periodic table.

A legitimate theory should therefore:

- predict structures that have not yet been observed but are structurally necessary;
- provide a principled pathway for evaluating those predictions.

A theory capable only of explaining known phenomena, but incapable of identifying structural vacancies, possesses explanatory value but insufficient generative power.

III. Applying the Framework to Superstring Theory

Within the structural framework, the central question is not whether eleven dimensions exist.

Instead, the relevant questions are:

1) Can the theory be reduced to the three fundamental structural roles?

2) Does it avoid uncontrolled landscape proliferation?

3) Does it possess a mechanism for predicting structural vacancies?

If these three principles are not satisfied, the structural status of the theory requires further evaluation.

This section deliberately draws no final conclusions. **Structural adjudication is not equivalent to theoretical rejection.**

IV. The Same Standards Must Apply to the Ultra-Metastructural Unification Theory

The same structural principles must also be applied to the author's own **Ultra-Metastructural Unification Theory.**

If the Structural Periodic Table is genuinely universal, no theory should receive special treatment.

4.1 Does It Achieve Minimal Generation?

The proposed master equation

$$dM/dt = \alpha_1 \nabla M + \alpha_2 I(E,S,C) + \alpha_3 Q(x)$$

must itself be examined. Can this formulation be further simplified? If so, it has not yet achieved structural minimality.

4.2 Does It Possess Unique Structural Closure?

Does the framework generate:

- multiple equivalent structures?
- unrestricted parameter adjustments?

If so, it would suffer from its own theoretical landscape and violate the principle of structural closure.

4.3 Can It Predict Structural Vacancies?

If the Structural Periodic Table is valid, it should predict:

- disciplines not yet established;
- higher structural levels not yet recognized;
- unified structural patterns not yet understood.

Without such predictive capability, the framework remains explanatory rather than generative.

V. A Fundamental Statement

If the Structural Periodic Table is valid, it protects no particular theory. The same structural criteria apply equally to:

- higher-dimensional string theories;
- the Ultra-Metastructural Unification Theory;
- every theoretical framework proposed by the present author.

Structure stands above theory. Every theory must conform to structural principles.

VI. A Criterion for Theoretical Maturity

If the Structural Periodic Table is capable of evaluating all higher-dimensional theories, it must also be capable of evaluating itself.

If it cannot evaluate itself, then it is merely another higher-dimensional theory.

VII. Final Position

This section does not reject higher-dimensional unified theories such as Superstring Theory.

Instead, it proposes that the legitimacy of higher-dimensional theories should not be judged by the number of dimensions they contain, but by the degree of structural closure they achieve.

The mission of the Structural Periodic Table is therefore not to replace existing theories, but to provide a universal language for structural adjudication.

Concluding Statement:

A genuinely unified theory is not the one that contains the greatest number of dimensions, but the one that can generate the broadest range of structures from the smallest set of fundamental structural principles.

Section 3. An Interpretive Framework for Unified Theories Based on the Point–Line–Circle Structure

A Minimal Structural Model for Unified Theory

Abstract:

The development of modern unified theories has generally followed two major directions. One seeks greater explanatory power by introducing additional theoretical degrees of freedom, such as higher-dimensional spaces, increasingly sophisticated symmetry groups, or new fundamental entities. The other attempts to construct broader mathematical frameworks capable of encompassing existing theories. Regardless of the approach, however, a more fundamental question remains:

What is the minimum structural framework required to generate a unified theory?

From the perspective of the Structural Periodic Table, this section proposes a minimal structural model consisting of three fundamental structural roles:

- **Point (•):** an irreducible generative source;
- **Line (|):** a mechanism of propagation and coupling;
- **Circle (O):** a mechanism of global closure and structural stability.

Together, these three elements constitute the minimal generative framework of a unified theory. This section does not attempt to construct a new physical theory. Rather, it proposes a structural language through which the common generative logic underlying possible unified theories may be described.

I. What Should a Unified Theory Actually Unify?

The Point–Line–Circle structural model proposed here is not intended to replace existing approaches to unification. Instead, it offers a higher-level structural language through which different unified theories may be interpreted.

Whether considering Superstring Theory, M-Theory, Loop Quantum Gravity, or other proposals, each may be viewed as a different organization of three fundamental structural processes:

- generation (Point),
- evolution (Line),
- closure (Circle).

Accordingly, this framework focuses not on specific mathematical implementations, but on the structural conditions that make unification possible.

Traditionally, unified theories seek to place different physical interactions within a common framework, including:

- the unification of electromagnetism and the weak interaction;
- the unification of electromagnetic, weak, and strong interactions;
- the unification of gravitation and quantum theory.

From the perspective of the Structural Periodic Table, however, unification means more than reducing the number of equations.

The deeper question is whether apparently different theories originate from the same underlying mechanism of structural generation.

If so, the objective of a unified theory is not to introduce additional entities, but to identify more fundamental structural principles.

II. The Point Structure: The Irreducible Generative Source

Every theory must ultimately answer one fundamental question: **Where does structure begin?**

Within the Structural Periodic Table, this role is represented by the **Point (•).** The Point represents:

- a local generative source;

- an irreducible structural origin;
- the initial condition from which all subsequent evolution proceeds.

Within existing theories, this role may correspond to:

- elementary particles;
- fundamental fields;
- vacuum fluctuations;
- elementary units of information.

The particular realization is less important than the underlying structural role.

Every unified theory requires some irreducible generative source. Without such an origin, no structure can emerge.

The Point therefore represents the necessary foundation of any unified theory.

III. The Line Structure: Propagation and Coupling

A generative source alone cannot produce an organized universe. Structures must also propagate, evolve, and interact. This function is represented by the **Line (|)**. The Line represents:

- interactions;
- coupling relationships;
- evolutionary pathways;
- information propagation.

Within existing theories, these mechanisms may appear as:

- electromagnetic propagation;
- quantum evolution;
- space-time dynamics;
- field couplings.

From the structural perspective, the Line transforms isolated points into organized systems.

Accordingly, the Line determines the dynamical content of a unified theory.

IV. The Circle Structure: Global Closure

If structures could expand indefinitely without constraint, no stable universe could emerge. A unified theory therefore requires a third structural component: the **Circle (O).**

The Circle represents:

- global closure;
- structural stability;
- self-consistency;
- conditions governing existence.

Within existing theories, these roles may appear as:

- conservation laws;
- boundary conditions;
- symmetry constraints;
- stable solution spaces.

From the perspective of the Structural Periodic Table, not every mathematically possible structure is physically realizable. Only structures satisfying appropriate closure conditions can persist over time.

The Circle therefore determines the conditions under which structures are able to exist.

V. A Minimal Structural Representation of Unified Theory

When the Point, Line, and Circle coexist, a minimal unified structure naturally emerges: **Point → Line → Circle**

or equivalently, **Generation → Evolution → Closure.**

This sequence constitutes the fundamental unit of structural periodicity.

Once structural closure is achieved, the completed structure becomes the starting point for a new cycle: **Point → Line → Circle → New Point**

This recursive process produces hierarchical organization.

Within the Structural Periodic Table, increasingly complex structures emerge through the repeated application of this generative cycle.

VI. Unified Theory from the Perspective of the Structural Periodic Table

Viewed in this way, competition among unified theories is no longer determined simply by:

- the number of dimensions;
- the number of elementary particles;
- the sophistication of mathematical formalism.

Instead, the more fundamental question becomes: **Which theory generates the greatest range of phenomena from the smallest structural foundation?**

Accordingly, the principal measure of a unified theory may not be its number of degrees of freedom, but its degree of **structural compression**.

If a theory continually requires additional structures merely to preserve internal consistency, its unifying power may remain incomplete. Conversely, a theory capable of explaining diverse phenomena through relatively few structural principles approaches the ideal of genuine unification.

VII. Conclusion

This section proposes an interpretive framework for unified theories based on the Structural Periodic Table.

Within this framework, every unified theory requires at least three fundamental structural roles:

- **Point (•):** the generative source;
- **Line (|):** the pathway of evolution;
- **Circle (O):** the mechanism of structural closure.

Together they constitute a minimal structural model for unified theory.

This section neither claims to have completed a unified theory nor rejects any existing proposal. Instead, it advances a structural hypothesis:

A genuine unified theory may not be the one possessing the greatest number of degrees of freedom, but the one capable of generating the widest range of phenomena from the smallest set of fundamental structures.

Concluding Statement:

A truly unified theory is not the one that contains the greatest number of dimensions, but the one that explains the greatest diversity of phenomena through the smallest set of fundamental structures.

Section 4. Case Studies in Structural Adjudication

A Structural Analysis of Superstring Theory, M-Theory, Loop Quantum Gravity, and the Ultra-Metastructural Theory

Abstract;

The previous sections introduced three principles of theoretical adjudication within the Structural Periodic Table:

1) the Principle of Minimal Generation;
2) the Principle of Unique Structural Closure;
3) the Principle of Predictive Structural Vacancies.

These principles are not intended to evaluate any particular theory in isolation. Rather, they constitute a structural evaluation framework that is independent of specific theoretical content.

This section illustrates the application of this framework to several representative approaches to unification, including Superstring Theory, M-Theory, Loop Quantum Gravity, and the author's proposed Ultra-Metastructural Theory.

The objective is not to determine which theory is correct or superior, but to examine their potential strengths and challenges from a structural perspective.

I. Review of the Structural Adjudication Principles

The Structural Periodic Table proposes three fundamental principles for evaluating unified theories.

1) Principle One: Minimal Generation

Can the theory generate a broad range of phenomena from the smallest possible set of fundamental structures?

2) Principle Two: Unique Structural Closure

Does the theory converge toward a stable and well-defined structural closure rather than an arbitrary collection of possible solutions?

3) Principle Three: Predictive Structural Vacancies

Can the theory identify structures that have not yet been observed but are structurally expected to exist?

Together, these principles provide a structural framework for evaluating unified theories.

II. Case A: Superstring Theory

2.1 Overview

Superstring Theory interprets elementary particles as different vibrational modes of fundamental strings while introducing a ten-dimensional space-time framework. Its principal strengths include:

- the unification of gauge interactions and gravitation;
- a highly coherent mathematical structure;
- extensive underlying symmetries.

2.2 Structural Analysis

1) Minimal Generation

Superstring Theory explains a wide variety of particles using a single fundamental object — the string.

From this perspective, it demonstrates considerable unifying potential. However, it also requires additional structures such as:

- ten-dimensional space-time;
- compactification mechanisms;
- Calabi–Yau manifolds.

Whether these additional structures are fundamentally irreducible remains an open question.

2) Unique Structural Closure

Superstring Theory is well known for the **Landscape Problem**, namely the existence of an enormous number of possible vacuum solutions.

Consequently, the uniqueness of its structural closure remains an important topic of ongoing research.

3) Predictive Structural Vacancies

The theory predicts phenomena such as:

- supersymmetric partner particles;
- higher-dimensional structures.

Many of these predictions, however, have yet to receive direct experimental confirmation.

2.3 Structural Assessment

Superstring Theory possesses significant unifying potential while continuing to face challenges concerning structural compression and the uniqueness of structural closure.

III. Case B: M-Theory

3.1 Overview

M-Theory is generally regarded as a higher-level framework unifying multiple versions of string theory.

Its central idea is that apparently different string theories may represent different aspects of a more fundamental structure.

3.2 Structural Analysis

1) Minimal Generation

M-Theory seeks to reduce fragmentation among competing string formulations.

From this perspective, it advances structural unification beyond individual string theories.

2) Unique Structural Closure

The theory still involves:

- multiple classes of solutions;
- the absence of a complete non-perturbative formulation.

Accordingly, questions concerning structural closure remain unresolved.

3) Predictive Structural Vacancies

M-Theory possesses considerable explanatory power but currently lacks a clearly established pathway for experimental verification.

3.3 Structural Assessment

M-Theory enhances conceptual unification, although questions concerning structural closure remain open.

IV. Case C: Loop Quantum Gravity

4.1 Overview

Loop Quantum Gravity seeks to quantize the structure of space-time itself.

Its central hypothesis is that space-time possesses an intrinsically discrete structure.

4.2 Structural Analysis

1) Minimal Generation

Loop Quantum Gravity avoids introducing additional spatial dimensions.

Its structural framework is therefore comparatively economical.

2) Unique Structural Closure

Its mathematical formulation is relatively well defined, although a complete dynamical formulation continues to be developed.

3) Predictive Structural Vacancies

The theory predicts phenomena including:

- quantized areas;
- quantized volumes.

These predictions provide potential avenues for empirical testing, although experimental verification remains challenging.

4.3 Structural Assessment

Loop Quantum Gravity exhibits a relatively compact structural formulation, although its scope of unification is more limited than some competing approaches.

V. Case D: The Ultra-Metastructural Theory

5.1 Overview

The Ultra-Metastructural Theory adopts the three structural roles — **Point (•), Line (|), and Circle (O)** — as its fundamental generative framework.

Its objective is not to construct a new elementary-particle theory directly, but to identify the common generative principles underlying diverse systems.

5.2 Structural Analysis

1) Minimal Generation

The theory attempts to interpret complex systems through three structural processes:

- Generation (Point);
- Evolution (Line);
- Closure (Circle).

From this perspective, it exhibits a strong tendency toward structural compression.

2) Unique Structural Closure

The framework requires every structure to achieve stable closure.
At present, however, its mathematical formalization and range of applicability require further development and rigorous analysis.

3) Predictive Structural Vacancies

If the Structural Periodic Table proves valid, the theory should be capable of predicting:

- structural levels not yet established;
- future disciplinary organizations;
- unified structural patterns not yet recognized.

Whether these predictions can be confirmed remains a subject for future investigation.

5.3 Structural Assessment

The Ultra-Metastructural Theory demonstrates a high degree of structural compression, while its predictive capability and mathematical rigor remain to be further developed.

VI. Comparative Summary

These four theories are selected not because they exhaust all approaches to unification, but because they represent four distinct structural strategies toward unification: higher-dimensional geometry, meta-unification, discrete quantum geometry, and minimal structural generation.

Theory	Minimal Generation	Unique Structural Closure	Predictive Structural Vacancies
Superstring Theory	Strong	Moderate	Moderate
M-Theory	Strong	Moderate	Moderate
Loop Quantum Gravity	Moderate	Relatively Strong	Moderate
Ultra-Metastructural Theory	Strong	To Be Evaluated	To Be Evaluated

Table 2-6-4-1: Comparison of the Ultra-Metastructural Theory with Representative Unified Theories from the Perspective of Structural Adjudication.

It should be emphasized that this comparison is **not intended as a ranking of competing theories.**

Rather, it illustrates how different theories exhibit different structural characteristics when evaluated under the same structural framework.

VII. Conclusion

The Structural Periodic Table is not intended to replace existing unified theories. Instead, it functions as a structural language through which unified theories may be examined according to common structural principles.

If the Structural Periodic Table is valid, every theory — including the author's own — must be evaluated according to exactly the same standards.

Structure belongs to no particular theory. Every theory must conform to structural principles.

Concluding Statement:

The purpose of structural adjudication is not to declare which theory is correct, but to ensure that every theory is evaluated according to the same structural standards.

Section 5. The Emergence of the Triadic Structural Pattern of Generation–Evolution–Closure

From Unified Theories to a Structural Language

Abstract:

The previous sections examined the structural boundaries of higher-dimensional unified theories, introduced the theoretical adjudication framework of the Structural Periodic Table, proposed the Point–Line–Circle structural model, and applied this framework to several representative approaches to unification.

This section advances a broader structural hypothesis: regardless of their specific mathematical formulations, theories capable of describing stable and persistent systems tend to exhibit a common triadic structural pattern consisting of **Generation, Evolution, and Closure**.

This pattern appears not only in mathematics and physics, but also in chemistry, life sciences, and complex systems. Consequently, the triadic structural pattern may represent not a discipline-specific principle, but a common structural language governing hierarchical organization across diverse domains.

I. What Should a Unified Theory Ultimately Unify?

Traditional unified theories primarily seek to unify:

- elementary particles;
- physical fields;
- fundamental interactions.

The Structural Periodic Table proposes a different question: **Do different theories share a common mechanism of structural generation?**

If such a mechanism exists, then the true object of unification may extend beyond physical laws themselves to the underlying processes through which structures emerge.

II. Generation: The Origin of Structure (Point •)

Every organized system requires a generative source.
Examples include:

- **Mathematics:** axioms and initial conditions;
- **Physics:** fundamental fields and vacuum fluctuations;
- **Chemistry:** chemical elements and elementary particles;
- **Life Sciences:** cells and genetic information.

Generation therefore represents the prerequisite for the emergence of structure.

Within the Structural Periodic Table, this role is represented by the **Point (•)**.

III. Evolution: The Development of Structure (Line |)

Once generated, a structure must propagate, develop, and interact. Examples include:

- **Mathematics:** recursion and algebraic operations;
- **Physics:** field propagation and interactions;
- **Chemistry:** reaction networks;
- **Life Sciences:** growth and heredity.

Evolution determines how structures expand and diversify.

Within the Structural Periodic Table, this process is represented by the **Line (|)**.

IV. Closure: The Condition for Structural Stability (Circle O)

If structures expanded indefinitely without constraint, no stable organization could emerge.

A mechanism of closure is therefore essential.

Examples include:

- **Mathematics:** proof completion and symmetry;
- **Physics:** conservation laws and stable modes;
- **Chemistry:** stable molecular configurations;
- **Life Sciences:** ecological cycles.

Closure determines whether a structure can persist over time.

Within the Structural Periodic Table, this role is represented by the **Circle (O).**

V. Cross-Disciplinary Correspondence of the Triadic Structure

Structural Process	Mathematics	Physics	Chemistry
Generation	Probability	Quantum Systems	Elements
Evolution	Algebra	Electromagnetism	Chemical Reactions
Closure	Geometry	Gravitation	Chemical Compounds

Figure 2-6-5-1: Manifestations of the Triadic Structural Pattern Across Different Scientific Levels.

From this perspective, the disciplines are not isolated domains.

They may instead represent different hierarchical manifestations of the same underlying structural pattern.

VI. From Unified Theories to a Structural Language

Superstring Theory and M-Theory seek to unify the laws of physics. The Structural Periodic Table seeks to unify the structural patterns underlying those laws. The two approaches therefore operate at different conceptual levels.

Unified theories ask: **"What is the universe made of?"**

The Structural Periodic Table asks:"**How are stable structures generated?"**

VII. A Structural Hypothesis

This section proposes the following structural hypothesis:

Every complex system capable of sustained existence must pass through three fundamental processes: **Generation → Evolution → Closure**

If any one of these processes is absent, the long-term persistence of the structure becomes impossible.

Accordingly, Generation, Evolution, and Closure may constitute the minimal structural cycle underlying organized systems.

VIII. Conclusion

From unified theories to complex systems, and from mathematics to the life sciences, numerous phenomena exhibit the recurring sequence: **Generation → Evolution → Closure.**

Accordingly, the Point–Line–Circle framework may represent not merely a descriptive language, but a cross-disciplinary pattern governing structural generation across hierarchical levels.

The Structural Periodic Table therefore seeks to explain not simply the relationships among disciplines, but why similar structural organizations repeatedly emerge across different levels of complexity.

Concluding Statement:

Unified theories seek to unify the laws of nature. The Structural Periodic Table seeks to unify the processes through which those laws emerge.

Scientific laws may vary from one domain to another, yet the underlying structural pattern may remain remarkably consistent. If this is true, what is ultimately universal is not any particular theory, but the generative structure shared by them all.

Section 6. The Structural Mapping Principle for Unified Theories

From Physical Theories to a Common Structural Language

Abstract:

The previous sections proposed that many complex systems naturally exhibit a triadic structural pattern consisting of **Generation, Evolution, and Closure**. If this pattern possesses genuine universality, then different scientific theories — despite employing distinct mathematical languages — may nevertheless share the same underlying logic of structural generation.

This section introduces the **Structural Mapping Principle**. Rather than attempting to derive the Point–Line–Circle (•–|–O) framework rigorously from any particular physical theory, the Structural Periodic Table seeks to identify the generative, evolutionary, and closure mechanisms that already exist within those theories.

If this principle is valid, diverse theoretical frameworks may be mapped onto a common structural coordinate system, thereby establishing a unified structural language that transcends individual mathematical formulations.

I. From Theoretical Unification to Structural Unification

Traditional unified theories generally seek to answer questions such as:

- How can different fundamental interactions be unified?
- How can different field equations be unified?
- How can different particle spectra be unified?

The Structural Periodic Table proposes a different perspective: **Do different theories share the same process of structural generation?**

If the answer is affirmative, then the true object of unification is not merely physical law, but structure itself.

Accordingly, the Structural Periodic Table does not require different theories to possess identical equations. Instead, it asks whether they can be mapped onto the same structural framework.

II. The Structural Mapping Principle

This work proposes the following Structural Mapping Principle:

Every mature scientific theory can be interpreted as consisting of three fundamental structural components.

1) Generative Structure (Point •)

This component answers: **Where does structure originate?**
Examples include:

- sources of mass,
- elementary particles,
- vacuum fluctuations,
- fundamental fields.

These serve as the generative origins within a theory.

2) Evolutionary Structure (Line |)

This component answers: **How does structure propagate?**
Examples include:

- fundamental interactions,
- field propagation,
- dynamical evolution,
- information transfer.

These constitute the evolutionary mechanisms of a theory.

3) Closure Structure (Circle O)

This component answers: **Why can structure remain stable?** Examples include:

- conservation laws,
- symmetry principles,
- boundary conditions,
- self-consistent solutions.

These provide the closure mechanisms required for persistent structural existence.

III. Structural Mapping of Newtonian Mechanics

Within Newtonian mechanics:

Structural Role	Physical Interpretation
Point (•)	Mass and initial conditions (*m*)
Line (I)	Equation of motion: F = ma
Circle (O)	Stable trajectories and orbital motion

Figure 2-6-6-1: Mapping of Newtonian Mechanics

Typical examples include planetary orbits and oscillatory systems. Thus Newtonian mechanics may be interpreted structurally as: **Mass Source → Dynamical Evolution → Stable Orbit**
or equivalently, **Point → Line → Circle**

IV. Structural Mapping of Maxwell's Theory

Within Maxwellian electrodynamics:

Structural Role	Physical Interpretation
Point (•)	Electric charge and current (ρ, J)
Line (|)	$\nabla \times \mathbf{E}, \nabla \times \mathbf{B}$
Circle (O)	Resonant modes and stable electromagnetic spectra

Figure 2-6-6-2: Mapping of Maxwell's Theory

Examples include resonant cavities, waveguides, and stable field modes. Thus electromagnetic theory likewise exhibits the structural sequence: **Generation → Evolution → Closure**

V. Structural Mapping of General Relativity

Within General Relativity:

Structural Role	Physical Interpretation
Point (•)	Energy–momentum distribution ($\mathbf{T}_{\{\mu\nu\}}$)
Line (|)	Spacetime curvature response $\mathbf{G}_{\{\mu\nu\}}$
Circle (O)	Global spacetime geometry ($\mathbf{g}_{\{\mu\nu\}}$)

Figure 2-6-6-3: Mapping of General Relativity

Examples include black-hole solutions and cosmological solutions. Accordingly, General Relativity may also be interpreted within the same structural framework.

VI. Structural Mapping of Quantum Mechanics

Within quantum theory:

Structural Role	Physical Interpretation
Point (•)	Local quantum fluctuations (Ψ_0)
Line (|)	Wave function evolution $i\hbar\, \partial\Psi/\partial t = H\Psi$
Circle (O)	Stable energy spectra and eigenstates

Figure 2-6-6-4: Mapping of Quantum Mechanics

Examples include atomic orbitals and quantum stationary states. Quantum mechanics therefore also demonstrates structural closure.

VII. Structural Mapping of Superstring Theory and M-Theory

7.1 The Hard Route: Formal Derivation

The fundamental ingredients of M-Theory include:

- eleven-dimensional spacetime,
- the supergravity action,
- *p*-branes,
- duality networks.

For example, its low-energy effective action is

$$S_{\{11\}} = \int d^{\{11\}}x\sqrt{\{-g\}}\left(R - \frac{\{1\}}{\{48\}}F_4^2\right) + \cdots \ldots$$

A rigorous derivation of the sequence: **Point → Line → Circle** would require successive analyses involving

- compactification,
- Kaluza–Klein reduction,

- vacuum selection,
- duality transformations,

ultimately leading to an effective four-dimensional theory.

Such a derivation would itself constitute an extensive research programme rather than a single structural argument.

7.2 The Structural Route

The Structural Periodic Table follows a different approach.

Rather than deriving structural roles from first principles, it seeks to identify structural roles already present within existing theories. For example:

1) Generation (Point)

Corresponds to

- fundamental branes,
- fundamental fields,
- vacuum states.

These function as **generative sources**.

2) Evolution (Line)

Corresponds to

- field propagation,
- brane interactions,
- duality transformations.

These function as **evolutionary mechanisms**.

3) Closure (Circle)

Corresponds to

- compactified geometries,
- Calabi–Yau constraints,

- consistency conditions.

These function as **closure mechanisms**. Accordingly, **Vacuum → Interaction → Consistency,** already possesses the structural outline: **Point → Line → Circle.**

7.3 Summary for Higher-Dimensional Unified Theories

Within higher-dimensional theories,

- **Point (•):** fundamental strings, branes, and vacuum states;
- **Line (|):** string vibrations, brane interactions, and duality transformations;
- **Circle (O):** compactified geometries, consistency conditions, and stable vacua.

Thus, despite their mathematical sophistication, these theories may still be interpreted as exhibiting the common sequence: **Generation → Evolution → Closure.**

VIII. A Unified Structural Coordinate System

If the Structural Mapping Principle is valid, then the following correspondence emerges:

Theory	**Point (•)**	**Line (|)**	**Circle (O)**
Newtonian Mechanics	Mass Source	Dynamics	Stable Orbit
Maxwell Theory	Electric Charge	Electromagnetic Propagation	Resonant Modes
General Relativity	Energy Source	Curvature Evolution	Spacetime Geometry
Quantum Mechanics	Quantum Fluctuation	Wave-function Evolution	Energy Spectrum

Superstring Theory	Strings / Branes	Interactions	Compactified Geometry
M-Theory	Fundamental Brane States	Duality Evolution	Consistent Vacuum

Table 2-6-6-5: Different theories employ different mathematical languages while sharing a common structural language.

Accordingly, although scientific theories utilize distinct mathematical formalisms, they may nevertheless share the same underlying structural organization.

IX. Conclusion

The Structural Periodic Table does not attempt to derive the Point–Line–Circle framework rigorously from M-Theory or any other unified theory.

Instead, it seeks to identify within existing theories their intrinsic

- generative mechanisms,
- evolutionary mechanisms,
- closure mechanisms.

If the Structural Mapping Principle proves valid, then Newtonian mechanics, Maxwell's theory, General Relativity, Quantum Mechanics, Superstring Theory, and M-Theory may all be interpreted within a common structural coordinate system.

Consequently, the Structural Periodic Table is concerned not with replacing existing theories, but with revealing the recurring patterns of structural generation that underlie them.

Concluding Statement:

Scientific theories may differ. Their equations may differ. Their dimensional frameworks may differ. Yet if the recurring processes of Generation, Evolution, and Closure remain the same, then what is fundamentally universal may not be the theories themselves, but the structural principles from which they emerge.

The Structural Periodic Table is not intended to replace existing scientific theories. Rather, it provides a common structural coordinate system through which different theories can be interpreted, compared, and connected.

References （Chapter 6）

Unified Theories and String Theory

[1] **Green, M. B., Schwarz, J. H., & Witten, E.** (1987). *Superstring Theory* (Vols. I–II). Cambridge University Press.

[2] **Polchinski, J.** (1998). *String Theory* (Vols. I–II). Cambridge University Press.

[3] **Becker, K., Becker, M., & Schwarz, J. H.** (2007). *String Theory and M-Theory: A Modern Introduction*. Cambridge University Press.

[4] **Witten, E.** (1995). *String Theory Dynamics in Various Dimensions*. *Nuclear Physics B*, **443**(1–2), 85–126.

[5] **Zwiebach, B.** (2009). *A First Course in String Theory* (2nd ed.). Cambridge University Press.

Quantum Gravity and Fundamental Physics

[6] **Einstein, A.** (1916). *The Foundation of the General Theory of Relativity*. *Annalen der Physik*, **49**, 769–822.

[7] **Dirac, P. A. M.** (1930). *The Principles of Quantum Mechanics*. Oxford University Press.

[8] **Weinberg, S.** (1995). *The Quantum Theory of Fields* (Vols. I–III). Cambridge University Press.

[9] **Rovelli, C.** (2004). *Quantum Gravity*. Cambridge University Press.

[10] **Rovelli, C., & Vidotto, F.** (2015). *Covariant Loop Quantum Gravity: An Elementary Introduction to Quantum Gravity and Spinfoam Theory*. Cambridge University Press.

[11] Ashtekar, A., & Lewandowski, J. (2004). *Background Independent Quantum Gravity: A Status Report. Classical and Quantum Gravity*, **21**, R53–R152.

Symmetry, Geometry and Mathematical Structure

[12] Weyl, H. (1952). *Symmetry*. Princeton University Press.

[13] Penrose, R. (2004). *The Road to Reality: A Complete Guide to the Laws of the Universe*. Jonathan Cape.

Complexity, Emergence and Structural Systems

[14] Anderson, P. W. (1972). *More Is Different. Science*, **177**(4047), 393–396.

[15] Simon, H. A. (1962). *The Architecture of Complexity. Proceedings of the American Philosophical Society*, **106**(6), 467–482.

[16] Holland, J. H. (1998). *Emergence: From Chaos to Order*. Oxford University Press.

[17] Prigogine, I., & Stengers, I. (1984). *Order Out of Chaos: Man's New Dialogue with Nature*. Bantam Books.

[18] Laughlin, R. B., & Pines, D. (2000). *The Theory of Everything. Proceedings of the National Academy of Sciences*, **97**(1), 28–31.

[19] Barabási, A.-L. (2016). *Network Science*. Cambridge University Press.

[20] Mitchell, M. (2009). *Complexity: A Guided Tour*. Oxford University Press.

Structural Periodicity and Metastructural Framework

[21] Chang, J. (2013). *Grand Ultimate Theory*. Universal Publishing.

[22] Chang, J. (2026). *Recursive Self-Organizing Evolutionary Breakthrough: Unifying Natural, Social and Life Sciences*. Universal Publishing.

[23] Chang, J. (2026). *Metastructural Unification (Physics Volume): Unifying Electromagnetism, Gravitation and Quantum Theory*. Universal Publishing.

Chapter Seven: Structural Periodicity in Physics

Figure 2–7: Crop circle illustrating Galactic Civilization. For details, see *Crop Circles*.

Chapter Overview:

This chapter examines five representative structural nodes in physics:

- **Particle Spectra** — microscopic structures
- **Field Modes** — propagation structures
- **Conservation Structures** — dynamical constraints
- **Phase Transition Hierarchies** — critical structural cycles
- **Cosmic Hierarchies** — macroscopic structures

Together these constitute five fundamental structural levels in physics and form a coherent chain of structural generation.

In summary, structural periodicity in physics may be expressed as: **Particle Spectra → Field Modes → Conservation Structures → Phase Transition Hierarchies → Cosmic Hierarchies**

Section 1. Fundamental Prototypes of Structural Periodicity in Physics — Natural Dynamics

Abstract:

In mathematics, structural periodicity appears through recursion, symmetry, periodic functions, moduli spaces, and spectral structures. In physics, however, structural periodicity emerges as stable organizational patterns produced through the interaction between natural dynamics and physical constraints.

Unlike pure mathematics, physical structures are governed not only by logical relationships but also by energy, interactions, and the geometry of space-time. Consequently, structural periodicity in physics is manifested as stable hierarchical organizations within natural dynamical systems.

From the perspective of the Structural Periodic Table, five representative structural prototypes can be identified in physics:

- particle spectra,
- field modes,
- conservation structures,
- phase-transition hierarchies,
- cosmic hierarchies.

Together they constitute the fundamental periodic framework underlying natural dynamical systems.

I. Particle Spectra: Discrete Hierarchies of Microscopic Structure

Modern physics demonstrates that the microscopic world is not continuous but consists of a series of discrete particle spectra. Representative examples include:

- lepton and quark spectra,
- hadron spectra,
- atomic energy spectra.

Within quantum theory these structures are typically represented by discrete eigenvalues, $\mathbf{E_n}$

The existence of particle spectra indicates that microscopic structures can exist only in specific stable states.

From the viewpoint of structural periodicity, particle spectra represent the microscopic manifestation of periodic structural organization.

II. Field Modes: Stable Patterns of Propagation

In continuous field theories, physical structures frequently appear as stable field modes. Typical examples include:

- electromagnetic modes,
- acoustic modes,
- quantum field excitations.

These modes satisfy specific boundary conditions and preserve their structural characteristics during propagation.

Mathematically, field modes are commonly represented by eigenfunctions of differential operators.

From the perspective of structural periodicity, field modes embody stable repeating patterns generated through structural propagation.

III. Conservation Structures: Stable Constraints in Dynamical Systems

Many periodic structures in physics arise from symmetry and conservation laws. Representative examples include:

- conservation of energy,
- conservation of momentum,
- conservation of electric charge.

According to Noether's theorem, deep connections exist between symmetry and conserved quantities.

These conservation laws provide stable constraints that preserve structural consistency throughout long-term evolution.

Within the language of structural periodicity, conservation laws may therefore be interpreted as the stable constraint structures governing natural dynamical systems.

IV. Phase Transition Hierarchies: Critical Cycles of Structural Reorganization

When system parameters reach critical values, structures may undergo abrupt reorganization—a phenomenon known as a phase transition. Typical examples include:

- solid–liquid–gas transitions,
- magnetic phase transitions,
- superconducting phase transitions.

Phase transitions demonstrate that physical systems may jump between distinct stable structural states.

From the viewpoint of structural periodicity, a phase transition represents simultaneously:

- the completion of one structural cycle, and
- the beginning of another.

Accordingly, phase-transition hierarchies constitute one of the principal mechanisms through which structural periodicity advances in natural systems.

V. Cosmic Hierarchies: Periodicity at Macroscopic Scales

At much larger scales, the Universe likewise exhibits pronounced hierarchical organization.

Representative examples include:

- planetary systems,
- stellar systems,

- galaxies,
- the large-scale cosmic web.

These hierarchical levels often display remarkably similar structural characteristics, including:

- gravitational aggregation,
- rotational organization,
- hierarchical distributions.

From the perspective of the Structural Periodic Table, these hierarchical organizations collectively represent the macroscopic manifestation of structural periodicity in nature.

VI. Summary

Through the five structural prototypes — particle spectra, field modes, conservation structures, phase-transition hierarchies, and cosmic hierarchies — we observe a common pathway of structural periodicity throughout physics: **Particle Spectra → Field Modes → Conservation Structures → Phase Transition Hierarchies → Cosmic Hierarchies**

This progression reveals the fundamental pattern governing structural evolution in natural systems.

In the following sections, we shall further examine how these physical structures correspond to the structural periodicity identified in mathematics, and how the same organizational principles continue into chemistry and, ultimately, the life sciences.

Section 2. A Triadic Interpretation of Structural Periodicity in Physics

Abstract:

The previous section introduced five representative prototypes of structural periodicity in physics: **particle spectra, field modes, conservation structures, phase-transition hierarchies, and cosmic hierarchies**. Although these structures belong to different branches of physics, from a deeper structural perspective they all reflect the interaction among three fundamental structural dynamics.

Within the structural language developed in this book, these three dynamics are abstracted as:

- **Point (•):** Local sources and structural generation
- **Line (|):** Propagation pathways and dynamical evolution
- **Circle (O):** Global constraints and stable closure

These three structures are not merely geometric metaphors, but abstract representations of three fundamental mechanisms governing natural systems.

Many seemingly unrelated physical phenomena can therefore be understood through the interaction of these three structural dynamics.

I. Point (•): Local Sources and Structural Generation

In physical systems, structure generally originates from localized sources. Typical examples include:

- the creation of particles;
- non-uniform energy density;
- field sources or mass sources.

These localized sources provide the starting point for structural generation.

In quantum theory, particles may be regarded as localized excitations of underlying fields. In gravitation, mass and energy density determine local space-time curvature. In electromagnetism, electric charges generate electromagnetic fields.

Within the language of structural periodicity, the **Point** therefore represents the local generative source and the origin of structural organization.

Without such localized sources, no dynamical structure can emerge.

II. Line (|): Propagation Pathways and Dynamical Evolution

Once local sources are established, structures propagate and evolve along characteristic pathways.

Within physical systems this propagation may appear as:

- wave propagation;
- field evolution;
- particle motion;
- energy transfer.

These processes are typically described by dynamical equations such as dM/dt. They characterize the temporal direction and evolutionary trajectory of natural systems.

From the viewpoint of structural periodicity, the **Line** represents the propagation and unfolding of structure through space and time. It forms the dynamical bridge connecting local generation with global organization.

III. Circle (O): Global Constraints and Stable Structure

When propagating structures establish stable relationships, structural closure emerges.

Within physical systems, such closure commonly appears as:

- conservation laws;
- boundary conditions;
- stable orbital configurations;

• discrete energy spectra.

These constraints prevent arbitrary evolution and restrict physical systems to a finite set of permitted structural configurations.

Representative examples include:

• electron orbitals in atoms;
• stable planetary orbits;
• standing-wave modes in wave systems.

These phenomena indicate that natural systems are governed by global constraints that guide structures toward stable closure.

Within the structural language developed here, this process is represented by the **Circle**.

IV. The Cyclic Relationship of the Triadic Structure

The three structural dynamics—Point, Line, and Circle—form a continuous cycle: **Point → Line → Circle → New Point**

This cycle provides a structural interpretation of periodicity in physics:

1) Local sources generate new structures (**Point**);
2) Structures propagate through dynamical pathways (**Line**);
3) Stable configurations emerge under global constraints (**Circle**);
4) Stable structures become the starting points for new hierarchical levels (**New Point**).

Examples include:

• particle spectra giving rise to new field structures;
• field modes producing new stable configurations;
• phase transitions generating new levels of material organization.

These are not isolated phenomena but different manifestations of the same triadic structural cycle across multiple physical scales.

V. The Triadic Structure Across the Three Physical Domains

Within electromagnetism, gravitation, and quantum theory, the triadic structural pattern appears in a remarkably consistent form.

Structural Element	Electro-magnetism	Gravitation	Quantum Theory
Point (•)	Charge source	Mass source	Local fluctuation
Line (|)	Field propagation	Space-time evolution	Wave function propagation
Circle (O)	Resonant modes	Geometric closure	Energy spectra

Table 2-7-2-1: Triadic structural correspondence among electromagnetism, gravitation, and quantum theory.

These correspondences suggest that natural dynamical systems exhibit similar structural cycles across different physical scales.

Although electromagnetism, gravitation, and quantum theory differ in mathematical formulation, they display an underlying structural isomorphism.

VI. Summary

Structural periodicity in physics may be understood as the recurring interaction among three fundamental structural dynamics:

- **Point:** Local generative sources;
- **Line:** Dynamical pathways of propagation;
- **Circle:** Stable structural closure.

Together they form the continuous cycle:

Generation → Evolution → Stabilization → New Hierarchical Generation

This cycle not only explains particle spectra, field modes, and phase-transition structures, but also provides a structural interpretation of hierarchical organization throughout the Universe.

In the following chapter, we shall examine how this same triadic structure appears in chemistry as the hierarchical relationship among **elements, reaction pathways, and molecular structures**, ultimately contributing to the interdisciplinary framework of the **Structural Periodic Table**.

Section 3. From Structural Periodicity in Physics to Structural Periodicity in the Natural Sciences

This chapter has examined structural periodicity from the perspective of physics, exploring its fundamental manifestations within natural dynamical systems. Through five representative structural prototypes — **particle spectra, field modes, conservation structures, phase-transition hierarchies, and cosmic hierarchies** — we observe the characteristic evolutionary pathway of structural periodicity in physical systems:

Particle Spectra → Field Modes → Conservation Constraints → Phase-Transition Reorganization → Cosmic Hierarchies

This progression demonstrates that, within physics, structures are generated not only through dynamical equations but are also organized into hierarchical forms through stable constraints.

Furthermore, this chapter has unified these phenomena using the language of the **triadic structural framework**. Within this framework:

- **Point (•)** represents local generative sources, such as particle excitations, energy density distributions, or local fluctuations;
- **Line (|)** represents the propagation and evolution of structure through space and time, including field propagation and dynamical trajectories;
- **Circle (O)** represents global constraints and stable configurations, including conservation laws, stable orbits, and discrete energy spectra.

These three structural elements form a continuous cycle: **Point → Line → Circle → New Point**

This cycle reveals the fundamental structural mechanism underlying natural dynamical systems. Local sources generate structure; structures evolve along dynamical pathways; stable patterns emerge under global constraints; and these stable configurations subsequently become the starting points for new hierarchical levels.

From this perspective, particle spectra, field modes, and phase-transition structures are not isolated physical phenomena, but different manifestations of the same triadic structural cycle operating across multiple scales.

Accordingly, structural periodicity in physics is not only an expression of natural dynamical systems but also serves as an important bridge toward understanding structural organization throughout the natural sciences. For example:

• In **chemistry**, elements, reaction pathways, and molecular structures form analogous hierarchical organizations.

• In **life sciences**, individual generation, evolutionary processes, and ecological stability likewise exhibit periodic structural patterns.

• In **social systems**, local events, developmental processes, and institutional structures display remarkably similar organizational dynamics.

These observations suggest that structural periodicity in physics is not an isolated phenomenon, but rather an essential intermediate layer within the broader framework of structural periodicity across the natural sciences: **Physics → Triadic Structure → Natural Sciences**

This chapter therefore completes an important conceptual transition: **Mathematical Structural Periodicity → Physical Structural Periodicity → Structural Periodicity in the Natural Sciences**

In the following chapter, we shall investigate structural periodicity within chemical systems, examining how the **periodic table of elements, reaction networks, and molecular structures** together constitute the hierarchical periodic organization of matter.

Note: Mathematics reveals the logical foundations of structural periodicity. Physics demonstrates its dynamical realization. Chemistry embodies its material manifestation. Together, they establish the three foundational pillars of the Structural Periodic Table.

References (Chapter 7)

The following references provide background for particle spectra, field theory, conservation laws, phase transitions, and large-scale cosmic structure. They situate the discussion within established physical theory.

[1] **Einstein, A.** (1916). The Foundation of the General Theory of Relativity. *Annalen der Physik*, 49, 769–822.

[2] **Dirac, P. A. M.** (1930). *The Principles of Quantum Mechanics*. Oxford: Oxford University Press.

[3] **Jackson, J. D.** (1999). *Classical Electrodynamics* (3rd ed.). Wiley.

[4] **Peskin, M. E. & Schroeder, D. V.** (1995). *An Introduction to Quantum Field Theory*. Westview Press.

[5] **Weinberg, S.** (1995). *The Quantum Theory of Fields*. Cambridge University Press.

[6] **Sachdev, S.** (2011). *Quantum Phase Transitions* (2nd ed.). Cambridge University Press.

[7] **Anderson, P. W.** (1972). More Is Different. *Science*, 177(4047), 393–396.

[8] **Laughlin, R. B. & Pines, D.** (2000). The Theory of Everything. *Proceedings of the National Academy of Sciences*, 97(1), 28–31.

[9] **Carroll, S. M.** (2019). *Spacetime and Geometry: An Introduction to General Relativity*. Cambridge University Press.

[10] **Chang, J.** (2026). *Recursive Self-Organizing Evolutionary Breakthrough — Unifying Natural, Social and Life Sciences*. Universal Publishing.

Chapter Eight: Structural Periodicity in Physics and Its Validation in Modern Physics

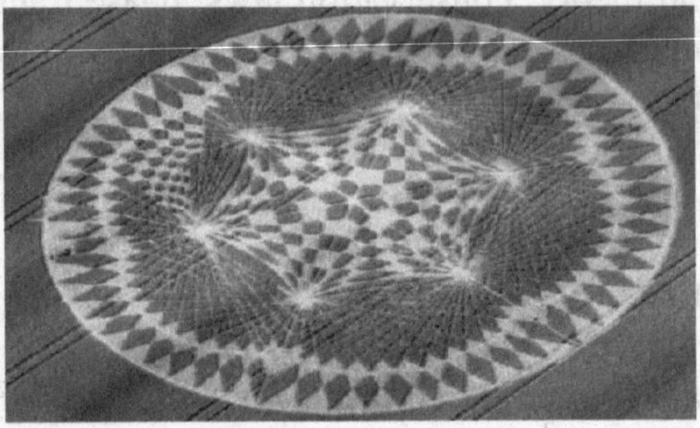

Figure 2-8: Crop circles illustrating a possible Galactic civilization. See 《Crop Circle》 for details.

Chapter Overview:

The development of modern physics has been characterized not only by the evolution of scientific theories, but also by the continuous evolution of the language used to describe nature.

From **Newtonian mechanics**, to **general relativity**, and eventually to **modern unified theories**, our description of the physical world has undergone several profound conceptual transitions — from the language of motion, to the language of geometry, and increasingly toward the language of structure.

This chapter therefore reexamines the development of modern physics from the perspective of the **Structural Periodic Table**, addressing a broader question: **Is modern physics gradually evolving from a language of motion into a language of structure?**

If this trend is indeed occurring, then the relationship among Newtonian mechanics, general relativity, and contemporary unified theories may represent more than successive changes in theoretical content. Instead, it may reflect a progressive evolution in the very way science describes the natural world.

Accordingly, this chapter is organized into four closely connected stages.

Section	Central Question	Structural Correspondence
Section 1	Why has the language of physics continually evolved?	Point (•)
Section 2	Why does general relativity already approach a structural theory?	Line (ǀ)
Section 3	What common structural framework underlies electro-magnetism, gravitation, and quantum theory?	Circle (O)
Section 4	Will future physics develop a structural-equation language?	Triadic Struc-tural Closure

Table 2-8-1: ***Four-stage evolution of structural language in modern physics.***

Through these discussions, this chapter argues that **Newton established the language of motion, Einstein established the language of geometry**, and that the continued development of unified theories may ultimately lead toward a **language of structure**.

This perspective is **not intended to replace existing physical theories**, but rather to provide a structural interpretation of the historical evolution of scientific thought. From the viewpoint of the Structural Periodic Table, **motion, geometry, and structure are not isolated paradigms, but successive stages in the continuous expansion of the descriptive language of natural science.**

Consequently, the primary objective of this chapter is not to advocate any particular physical theory, but to explore **how modern physics has evolved from equations of motion, to geometric equations, and ultimately toward structural equations.**

In summary, **physics validates whether structural principles possess genuine correspondence with nature (Physical Correspondence).**

Section 1. From Equations of Motion to Structural Equations – Three Linguistic Transitions in Modern Physics

Abstract:

The evolution of modern physics is reflected not only in the expansion of theoretical knowledge, but also in the continual evolution of its descriptive language. From the **equations of motion** of Newtonian mechanics, to the **geometric equations** of Einstein's general relativity, and more recently to the emerging **structural language** found in complexity science, information science, and unified theories, physics appears to be undergoing a deeper transformation in the way it describes nature.

From the perspective of the **Structural Periodic Table**, this chapter proposes a new interpretive framework in which the historical development of modern physics may be understood as a sequence of three linguistic transitions: **Motion → Geometry → Structure.**

Equations of motion describe **how objects change**; geometric equations describe **how relationships are organized**; structural equations seek to describe **how physical regularities themselves are generated**. This section reexamines the evolution of modern physics through this structural perspective.

I. Physics Evolves Not Only Through Theory but Also Through Language

Scientific progress is often understood as the discovery of new particles, the formulation of new equations, or the development of new theories. From a broader perspective, however, the history of physics is equally a history of the evolution of its descriptive language.

Different historical periods have focused on fundamentally different questions:

- **Newtonian physics:** *How do objects move?*

- **Einsteinian physics:** *How is spacetime curved?*
- **Contemporary unified theories:** *How are structures generated?*

Accordingly, the development of physics represents not merely an accumulation of knowledge, but also a succession of increasingly expressive conceptual languages.

II. The First Transition: The Era of Equations of Motion

Modern physics began with the description of motion.
Newton's Second Law,

F = ma

became the central equation of classical mechanics. Within this framework:

- **mass** represents the physical object;
- **force** acts as the causal agent;
- **acceleration** is the observable consequence.

The theory is fundamentally concerned with explaining **how physical objects move**.
The era of Newtonian mechanics may therefore be regarded as the **Era of Equations of Motion**, characterized by:

- clearly defined physical objects,
- direct causal relationships,
- descriptions of local dynamical processes.

Motion thus became the primary language through which nature was understood.

III. The Second Transition: The Era of Geometric Equations

At the beginning of the twentieth century, **general relativity** fundamentally transformed this picture.

Einstein's field equation,

$$G_{\{\mu\nu\}} = \frac{\{8\pi G\}}{\{c^4\}} T_{\{\mu\nu\}}$$

no longer describes force directly. Instead, it relates:

- mass and energy,
- spacetime curvature,
- global geometric structure.

Gravity is reinterpreted as the geometric response of spacetime itself. The principal object of study is therefore no longer the trajectory of individual particles, but the geometry of spacetime as a whole.

General relativity thus marks the beginning of the **Era of Geometric Equations**, characterized by:

- emphasis on global relationships,
- geometric consistency,
- the description of nature through spacetime geometry.

IV. The Third Transition: The Era of Structural Equations

With the emergence of quantum theory, complexity science, and modern unified theories, physics increasingly confronts new questions:

- Why do certain structures emerge?
- Why do particular regularities repeatedly appear?
- Why do different systems exhibit remarkably similar organizational patterns?

These questions extend beyond the traditional domains of motion and geometry. Instead, they concern the **mechanisms of structural generation**.

From the perspective of the Structural Periodic Table, future physics may gradually shift from **describing motion** toward **describing structure**. This motivates the introduction of **structural equations** as a new descriptive framework.

Their purpose is not merely to predict dynamical evolution or characterize geometric relationships, but to investigate **how structures are generated, evolve, and become stabilized.**

V. The Basic Idea of Structural Equations

In this work, structural evolution is represented by the following master equation:

$$\mathbf{dM/dt = \alpha_1 \nabla M + \alpha_2 I(E,S,C) + \alpha_3 Q(x)}$$

Here,

- **∇M** represents **local generation and fluctuations**;
- **I(E,S,C)** represents **global feedback and structural constraints**;
- **Q(x)** represents **propagation, coupling, and external driving**.

Within the Structural Periodic Table, these correspond respectively to:

- **Point (•):** generation,
- **Circle (O):** structural closure,
- **Line (|):** evolution.

Together they form the recurring structural cycle: **Generation → Evolution → Closure.**

VI. The Hierarchy of Motion, Geometry, and Structure

Viewed historically, the three descriptive languages form a natural hierarchy:

1) Motion — describing how objects change;
2) Geometry — describing how relationships are organized;
3) Structure — describing how physical regularities are generated.

These three perspectives should not be regarded as competing alternatives. Rather, each represents a progressively higher level of scientific description: **Motion → Geometry → Structure**

- Motion emphasizes local dynamics.
- Geometry emphasizes relational organization.
- Structure emphasizes generative mechanisms.

VII. The Place of Physics Within the Structural Periodic Table

From the viewpoint of the Structural Periodic Table, the historical development of physics itself exhibits a hierarchical periodic progression.

- Newtonian mechanics emphasizes **motion**.
- General relativity emphasizes **geometry**.
- Modern unified theories increasingly emphasize **structure**.

Accordingly, the evolution of modern physics may be interpreted not only as the continual refinement of theories, but also as the progressive evolution of scientific language itself.

If this trend continues, the principal object of future unified theories may no longer be particles, fields, or even spacetime, but **structure itself**.

VIII. Conclusion

The development of modern physics may be viewed as three successive linguistic transitions:

• **First Transition:** Equations of Motion → the study of motion;

• **Second Transition:** Geometric Equations → the study of relationships;

• **Third Transition:** Structural Equations → the study of generative mechanisms.

From the perspective of the Structural Periodic Table, motion, geometry, and structure are not mutually exclusive paradigms, but successive stages in the continuing evolution of scientific description.

Future unified theories may therefore seek not only to unify the laws of physics themselves, but also the structural mechanisms from which those laws emerge.

Concluding Statement:

Newton unified motion. Einstein unified geometry. A future unified theory may ultimately seek to unify structure itself.

Section 2. General Relativity as an Emerging Structural Theory — From Geometric Dynamics to Structural Dynamics

Abstract:

General relativity is widely regarded as the ultimate geometric formulation of gravitation. Unlike Newtonian mechanics, which explains motion through forces, Einstein reinterpreted gravity as the curvature of spacetime itself. From the perspective of the **Structural Periodic Table**, however, the significance of general relativity extends beyond the geometrization of gravity. More fundamentally, it represents a shift away from traditional dynamical descriptions toward the description of global structural organization.

This section proposes that although general relativity remains fundamentally a geometric theory, its underlying conceptual framework already approaches that of a structural theory. Mass-energy, spacetime curvature, and geometric configuration together form a structural chain of **generation → evolution → closure**, making general relativity an important bridge between classical dynamics and a future theory of structural organization.

I. Newtonian Mechanics Studies Motion

The central question of classical mechanics is straightforward: **How do physical objects move?**

Newton's Second Law, $F = ma$, describes:

- the action of force,
- the generation of acceleration,
- the resulting change in motion.

Its principal object of study is therefore the **state of motion**.

Accordingly, classical mechanics represents a typical **dynamical theory**, whose primary objective is to describe local physical processes.

II. General Relativity Studies Relationships

Einstein's field equation,

$$G_{\{\mu\nu\}} = \frac{\{8\pi G\}}{\{c^4\}} T_{\{\mu\nu\}}$$

no longer describes force directly. Instead, it relates:

- the distribution of mass and energy,
- the response of spacetime,
- the emergence of geometric structure.

General relativity therefore studies not motion itself, but the **relationships** that organize spacetime.

Gravity is no longer interpreted as a force acting between objects, but as the global organization of spacetime geometry.

III. The Structural Chain Within General Relativity

Viewed through the perspective of the Structural Periodic Table, general relativity already contains a complete structural process: **Generation → Evolution → Closure**

3.1 Generation: Sources of Mass and Energy

The stress-energy tensor, $T_{\{\mu\nu\}}$ represents:

- mass,
- energy,
- momentum.

These quantities constitute the generative source of spacetime structure. Within the structural framework, this corresponds to the **Point (•)** — the local source from which structure originates.

3.2 Evolution: The Propagation of Curvature

The Einstein tensor, $G_{\{\mu\nu\}}$

describes how local sources modify the surrounding geometry.

Curvature propagates throughout spacetime, continuously influencing its global configuration.

This corresponds to the **Line (|)**—the evolutionary process of structural development.

3.3 Closure: Global Spacetime Geometry

The metric tensor, $\mathbf{g}_{\{\mu\nu\}}$

determines the global geometric organization of spacetime, including:

- black-hole solutions,
- cosmological solutions,
- stable geometric configurations.

This corresponds to the **Circle (O)** — the mechanism of structural closure.

General relativity therefore naturally forms the structural chain: **Mass-Energy Source → Curvature Evolution → Geometric Closure**

IV. Why Does General Relativity Already Approach Structural Theory?

Traditional dynamics typically follows a simple causal sequence: **Cause → Effect**

For example, $\mathbf{F} \rightarrow \mathbf{a}$.

General relativity introduces a fundamentally different viewpoint: **Local Sources → Global Structure**

represented schematically by $\mathbf{T}_{\{\mu\nu\}} \rightarrow \mathbf{g}_{\{\mu\nu\}}$

This perspective extends beyond conventional dynamics because the global geometry subsequently influences local motion.

A feedback cycle emerges: **Local Sources → Global Geometry → Local Motion**

Such reciprocal organization is a defining characteristic of structural systems.

V. General Relativity Has Not Yet Become a Complete Structural Theory

Despite these advances, general relativity remains primarily a theory of **spacetime geometry**, rather than a theory of **structural generation**. It successfully explains:

- why spacetime curves,
- how geometry evolves,

but it does not explicitly address a deeper question: **Why do these structural organizations arise in the first place?**

From this perspective, general relativity approaches structural theory without fully becoming one.

VI. From General Relativity to Structural Equations

From the viewpoint of the Structural Periodic Table, future unified theories may continue this conceptual progression:

- Newton investigated **motion**.
- Einstein investigated **geometry**.
- Structural theory investigates **mechanisms of generation**.

General relativity may therefore be interpreted as the first major transition from classical dynamics toward structural science.

VII. Reinterpreting General Relativity Within the Structural Periodic Table

Within the Structural Periodic Table, general relativity is viewed not merely as a gravitational theory, but as a theory of structural formation.

Its fundamental chain can be expressed as: **Point (•) → Line (|) → Circle (O)**

or equivalently, **Mass-Energy Source → Curvature Propagation → Geometric Closure**

Seen from this perspective, general relativity already exhibits the essential characteristics of structural periodicity.

VIII. Conclusion

The greatest contribution of general relativity lies not only in geometrizing gravitation, but also in transforming physics from the study of local dynamics to the study of global structural organization.

For the first time, physics began to investigate:

- global relationships,
- structural consistency,
- geometric closure.

General relativity therefore remains a geometric theory, yet it already displays many of the defining features of a structural theory. It may be regarded both as the culmination of the geometric era and as a possible point of departure toward a future structural era.

Concluding Statement:

Newton revealed how objects move. Einstein revealed how spacetime is organized. Structural theory seeks to answer a deeper question: why such modes of organization can exist at all.

Viewed in this way, Newton established the language of dynamics, Einstein established the language of geometry, and twenty-first-century science may gradually be moving toward a language capable of describing the mechanisms of structural generation.

Section 3. The Common Structural Framework of Electromagnetism, Gravitation, and Quantum Theory – From Three Theories to a Unified Structural Language

Abstract:

Modern physics is primarily built upon three foundational theoretical frameworks: **electromagnetic theory, general relativity, and quantum theory**. These theories have successfully explained electromagnetic phenomena, gravitation, and the microscopic world, respectively. Nevertheless, each employs its own mathematical formalism, making their unification one of the central challenges of contemporary physics.

Rather than attempting to unify these theories mathematically, this section approaches the problem from the perspective of the **Structural Periodic Table**, asking whether they share a common structural framework. We argue that each theory contains an identifiable structural sequence of **generation → evolution → closure**. If this structural correspondence is valid, then although their mathematical languages differ, the three theories may nevertheless be interpreted through a common structural language.

I. The Problem of Unification and the Problem of Structure

Traditional unified theories focus on questions such as:

- the unification of electromagnetism and the weak interaction;
- the unification of gravitation and quantum theory;
- grand unified theories and string theory.

These approaches primarily ask: **How can the equations be unified?**

The Structural Periodic Table proposes a different question: **Can the underlying structures be unified?**

In other words, **Do different physical theories share the same generative mechanism?**

II. The Structural Framework of Electromagnetic Theory

Maxwell's equations established the foundations of classical electromagnetism. Viewed structurally, the theory may be interpreted as follows.

2.1 Generation

Electric charge and electric current, **ρ; J**
serve as the sources of electromagnetic fields.
They correspond to the **Point (•)**—the local source of structural generation.

2.2 Evolution

The electromagnetic fields, **E; B,**
propagate through space and interact with matter.
This corresponds to the **Line (|)**—the process of propagation and coupling.

2.3 Closure

Stable resonant configurations, including **laser cavities, resonators, waveguides**, represent stable electromagnetic structures.
These correspond to the **Circle (O)** — the mechanism of structural closure.
Electromagnetic theory may therefore be represented schematically as: **Charge → Propagation → Resonance**

III. The Structural Framework of Gravitation

General relativity provides the modern theory of gravitation. Within the structural framework:

3.1 Generation

The stress-energy tensor, $\mathbf{T_{\{\mu\nu\}}}$ represents mass-energy sources. This corresponds to the **Point (•)**.

3.2 Evolution

The Einstein tensor, $\mathbf{G_{\{\mu\nu\}}}$ describes the evolution of spacetime curvature.This corresponds to the **Line (|)**.

3.3 Closure

The metric tensor, $\mathbf{g_{\{\mu\nu\}}}$ determines the global geometric organization of spacetime, including

- black holes,
- galactic structures,
- cosmological solutions.

This corresponds to the **Circle (O)**.

Accordingly, gravitation may be viewed structurally as: **Mass-Energy → Curvature → Geometry**

IV. The Structural Framework of Quantum Theory

Quantum theory describes microscopic physical systems. From the structural perspective:

4.1 Generation

Quantum fluctuations, $\mathbf{\Psi_0}$
serve as local probabilistic sources. These correspond to the **Point (•)**.

4.2 Evolution

The evolution of the wavefunction, $i\hbar\, \partial\Psi/\partial t = H\Psi$
represents the dynamical development of quantum states.

This corresponds to the **Line (|)**.

4.3 Closure

Stable quantum structures include

- atomic orbitals,
- eigenstates,
- discrete energy spectra.

These correspond to the **Circle (O)**.

Quantum theory may therefore be summarized structurally as:
Fluctuation → Evolution → Energy Spectrum

V. A Common Structural Framework for the Three Theories

Placing these theories within a common structural coordinate system yields:

Structural Stage	**Electromagnetism**	**Gravitation**	**Quantum Theory**
Point (•): Generation	Charge	Mass-Energy Source	Quantum Fluctuation
Line (|): Evolution	Electromagnetic Propagation	Curvature Evolution	Wavefunction Evolution
Circle (O): Closure	Resonant Modes	Spacetime Geometry	Energy Spectrum

Table 2-8-3-1: A common structural framework underlying the three major physical theories.

Although their mathematical formulations differ substantially, their structural roles exhibit remarkable similarities.

Furthermore, each theory emphasizes a different aspect of the overall structural cycle:

Theory	**Primary Structural Emphasis**	**Structural Role**
Quantum Theory	Mechanisms of generation	**Point (•)**
Electromagnetic Theory	Mechanisms of propagation	**Line (|)**
Gravitation	Mechanisms of structural closure	**Circle (O)**

Table 2-8-3-2: Generation, evolution, and closure in the three major physical theories.

From the conventional perspective, electromagnetism, gravitation, and quantum mechanics are regarded as three distinct theories.

From the perspective of the Structural Periodic Table, however, they emphasize three complementary stages of a single structural process.

VI. Structural Framework Versus Unified Equations

Traditional unified theories seek: **one universal equation.**

The Structural Periodic Table instead seeks: **one universal structural language.**

Within this language, electromagnetism, gravitation, and quantum theory are not isolated disciplines but different manifestations of the same structural process operating at different organizational levels.

Accordingly, what is unified may not necessarily be the equations themselves, but the underlying structural pattern.

VII. A Structural Reinterpretation

From the viewpoint of the Structural Periodic Table:

- **Electromagnetism** primarily emphasizes **propagation**.
- **Gravitation** primarily emphasizes **closure**.
- **Quantum theory** primarily emphasizes **generation**.

Together they form the complete structural cycle: **Point (•) → Line (|) → Circle (O)**
or equivalently, **Generation → Evolution → Closure**
The three theories may therefore be interpreted not as isolated descriptions of nature, but as complementary expressions of one continuous structural process.

VIII. Conclusion

Although electromagnetic theory, gravitation, and quantum theory employ different mathematical languages, each exhibits the same underlying structural sequence: **Generation → Evolution → Closure**
The Structural Periodic Table therefore does not seek to replace existing physical theories with a single equation. Instead, it proposes a higher-level structural coordinate system within which these theories can be interpreted coherently.
Within this framework, electromagnetism, gravitation, and quantum theory become not isolated theoretical systems, but different expressions of a common structural language.

Concluding Statement:

Electromagnetism reveals how structure propagates.
Gravitation reveals how structure achieves closure.
Quantum theory reveals how structure is generated.
The Structural Periodic Table suggests that these three processes have always belonged to the same structural cycle.

Section 4. Structural Equations as a Candidate Language for Future Physics — From Motion and Geometry to the Mechanisms of Structural Generation

Abstract:

The preceding sections have shown that the historical development of modern physics has progressed from **equations of motion** to **geometric equations**, while electromagnetic theory, gravitation, and quantum theory all exhibit a common structural framework characterized by **generation, evolution, and closure**.

This section proposes a further possibility: the fundamental language of future physical theory may evolve beyond descriptions of motion or geometry toward descriptions of **structure**. The objective of a structural equation is not to replace existing physical theories but to provide a unified structural language capable of simultaneously describing **generation, evolution, and closure**. By connecting the ternary structural framework of the Structural Periodic Table with a unified structural equation, this chapter explores its potential significance as a candidate language for future theoretical physics.

I. Why Has Physics Continuously Changed Its Language?

The history of physics may be viewed as a sequence of successive changes in scientific language.

1.1 The Newtonian Era

The central language was the equation of motion,

$F = ma$,

whose primary question was: **How do objects move?**

1.2 The Einsteinian Era

The principal language became the geometric field equation,

$$G_{\{\mu\nu\}} = \frac{\{8\pi G\}}{\{c^4\}} T_{\{\mu\nu\}}$$

whose central question was: **How is spacetime organized?**

Contemporary unified theories increasingly confront problems involving

- multiscale systems,
- complex feedback,
- self-organizing structures.

Such problems extend beyond the traditional frameworks of motion and geometry.

Consequently, a new structural language may gradually emerge.

II. What Is a Structural Equation?

Equations of motion describe **changes of state**.

Geometric equations describe **relations among structures**.

Structural equations seek to describe **mechanisms of generation**.

They ask questions such as:

- Why do particular structures emerge?
- Why do certain regularities persist?
- Why do similar structural patterns repeatedly appear in different systems?

Accordingly, structural equations focus on the organization of structure itself.

III. A Unified Structural Representation in the Structural Periodic Table

This book adopts the following structural representation:

$$\mathbf{dM/dt = \alpha_1 \nabla M + \alpha_2 I(E,S,C) + \alpha_3 Q(x)}$$

Here,

- M denotes the structural state of the system;
- $\boldsymbol{\alpha_1 \nabla M}$ represents local generation and fluctuations, corresponding to the **Point (•)**;
- $\boldsymbol{\alpha_2}$$\mathbf{I(E,S,C)}$ represents global feedback and structural constraints, corresponding to the **Circle (O)**;
- $\boldsymbol{\alpha_3}$$Q(x)$ represents propagation and coupling, corresponding to the **Line (|)**.

The structural equation therefore attempts to describe simultaneously the three fundamental processes of **Generation → Evolution → Closure,** within a single mathematical framework.

IV. Structural Mapping of Electromagnetism, Gravitation, and Quantum Theory

From the perspective of the Structural Periodic Table,

1) Quantum theory primarily emphasizes local generation, corresponding to $\mathbf{\nabla M}$

2) Gravitational theory emphasizes global constraints and geometric closure, corresponding to **I(E,S,C);**

3) Electromagnetic theory emphasizes propagation and coupling, corresponding to **Q(x).**

This yields the structural representation:

dM/dt = $\boldsymbol{\alpha_1}$**{Quantum}+** $\boldsymbol{\alpha_2}$**{Gravitation}+**$\boldsymbol{\alpha_3}$**{Electromagnetism}.**

This expression is **not intended as a rigorous physical derivation**, but rather as a **structural mapping** illustrating how different physical theories may occupy complementary roles within a unified structural framework.

V. Structural Equations Versus Traditional Unified Theories

Conventional unified theories generally pursue **a final physical equation**, such as

- Grand Unified Theories (GUTs),
- String Theory,
- M-Theory.

Their objective is to unify the fundamental interactions.

Structural equations pursue a different goal.

Rather than replacing existing theories, they seek to identify the structural principles shared by diverse theoretical frameworks.

Accordingly, the structural equation should be viewed as a higher-level structural language rather than as a replacement for established physical theories.

VI. Potential Advantages of Structural Equations

If the proposed structural language proves useful, it may possess several notable characteristics.

6.1 Cross-disciplinary Applicability

The same structural framework may extend beyond physics to

- mathematics,
- chemistry,
- life sciences,
- social systems.

6.2 Multiscale Integration

The structural equation may simultaneously describe

- microscopic generation,
- intermediate-scale propagation,
- macroscopic closure.

6.3 Interpretation of Structural Periodicity

It may also provide an explanation for why similar structural patterns repeatedly emerge across different hierarchical levels.

VII. Limitations of the Structural Equation

It is important to emphasize that the structural equation proposed here is **not claimed to be a new fundamental law of physics**.

At its present stage, it should be regarded as **a structural model,** rather than **an experimentally established physical law**.

Its primary purpose is to provide a possible conceptual framework for structural unification. Further progress will require

- greater mathematical rigor,
- correspondence with experimental observations,
- independent theoretical verification.

VIII. Conclusion

Modern physics has progressed through

- the era of motion equations,
- the era of geometric equations,

and may gradually enter **the era of structural equations**.

Structural equations do not seek to replace electromagnetic theory, general relativity, or quantum mechanics. Instead, they attempt to identify the structural principles shared by these theories.

If future unified theories succeed not only in unifying physical laws but also in unifying the mechanisms through which those laws emerge, then structural language may become a new form of theoretical description.

Concluding Statement: Newton investigated motion. Einstein investigated geometry. Future physics may ultimately investigate why structure itself can be generated.

References (Chapter 8):

The references for this chapter are organized to follow the historical and conceptual progression of modern physics:
Motion → Geometry → Electromagnetism → Quantum Theory → Complex Systems → Structural Theory

This sequence corresponds directly to the organization of the chapter:

- **Section 1:** Equations of Motion
- **Section 2:** General Relativity
- **Section 3:** The Structural Correspondence of Electromagnetism, Gravitation, and Quantum Theory
- **Section 4:** Structural Equations as a Candidate Theoretical Language

Together they provide a coherent reference framework tracing the evolution of modern physics from classical mechanics toward structural approaches to scientific theory.

Classical Mechanics and Dynamical Descriptions

[1] **Newton, I.** (1687). *Philosophiæ Naturalis Principia Mathematica*. Royal Society.

[2] **Goldstein, H., Poole, C., & Safko, J.** (2002). *Classical Mechanics* (3rd ed.). Addison-Wesley.

[3] **Lanczos, C.** (1986). *The Variational Principles of Mechanics* (4th ed.). Dover Publications.

Relativity, Gravitation and Geometrical Descriptions

[4] Einstein, A. (1916). The Foundation of the General Theory of Relativity. *Annalen der Physik*, 49, 769–822.

[5] Einstein, A. (1923). *The Meaning of Relativity*. Princeton University Press.

[6] Misner, C. W., Thorne, K. S., & Wheeler, J. A. (1973). *Gravitation*. W. H. Freeman.

[7] Hawking, S. W., & Ellis, G. F. R. (1973). *The Large Scale Structure of Space-Time*. Cambridge University Press.

[8] Wald, R. M. (1984). *General Relativity*. University of Chicago Press.

[9] Penrose, R. (2004). *The Road to Reality: A Complete Guide to the Laws of the Universe*. Jonathan Cape.

Electromagnetic Theory

[10] Maxwell, J. C. (1873). *A Treatise on Electricity and Magnetism*. Oxford University Press.

[11] Jackson, J. D. (1999). *Classical Electrodynamics* (3rd ed.). John Wiley & Sons.

[12] Griffiths, D. J. (2017). *Introduction to Electrodynamics* (4th ed.). Cambridge University Press.

Quantum Theory and Unified Theories

[13] Dirac, P. A. M. (1930). *The Principles of Quantum Mechanics*. Oxford University Press.

[14] von Neumann, J. (1955). *Mathematical Foundations of Quantum Mechanics*. Princeton University Press.

[15] Feynman, R. P., Leighton, R. B., & Sands, M. (1965). *The Feynman Lectures on Physics, Volume III: Quantum Mechanics*. Addison-Wesley.

[16] Weinberg, S. (1995). *The Quantum Theory of Fields* (Vols. I–III). Cambridge University Press.

[17] Witten, E. (1995). String Theory Dynamics in Various Dimensions. *Nuclear Physics B*, 443(1–2), 85–126.

[18] Green, M. B., Schwarz, J. H., & Witten, E. (1987). *Superstring Theory* (Vols. I–II). Cambridge University Press.

[19] Polchinski, J. (1998). *String Theory* (Vols. I–II). Cambridge University Press.

[20] Rovelli, C. (2004). *Quantum Gravity*. Cambridge University Press.

[21] Sakurai, J. J., & Napolitano, J. (2017). *Modern Quantum Mechanics* (2nd ed.). Cambridge University Press.

Complexity, Emergence and Structural Descriptions

[22] Anderson, P. W. (1972). More Is Different. *Science*, 177(4047), 393–396.

[23] Simon, H. A. (1962). The Architecture of Complexity. *Proceedings of the American Philosophical Society*, 106(6), 467–482.

[24] Prigogine, I., & Stengers, I. (1984). *Order Out of Chaos*. Bantam Books.

[25] Holland, J. H. (1998). *Emergence: From Chaos to Order*. Oxford University Press.

[26] Mitchell, M. (2009). *Complexity: A Guided Tour*. Oxford University Press.

[27] Barabási, A.-L. (2016). *Network Science*. Cambridge University Press.

Scientific Paradigms and Structural Frameworks

[28] Kuhn, T. S. (1962). *The Structure of Scientific Revolutions*. University of Chicago Press.

[29] Weyl, H. (1952). *Symmetry*. Princeton University Press.

[30] Laughlin, R. B., & Pines, D. (2000). The Theory of Everything. *Proceedings of the National Academy of Sciences*, 97(1), 28–31.

Structural Periodicity and Metastructural Framework

[31] Chang, J. (2013). *Great Ultimate Theory*. Universal Publishing.

[32] Chang, J. (2026). *Recursive Self-Organizing Evolutionary Breakthrough: Unifying Natural, Social and Life Sciences*. Universal Publishing.

[33] Chang, J. (2026). *Metastructural Unification (Physics Volume): Unifying Electromagnetism, Gravitation and Quantum Theory*. Universal Publishing.

Chapter Nine: Structural Periodicity in Chemistry

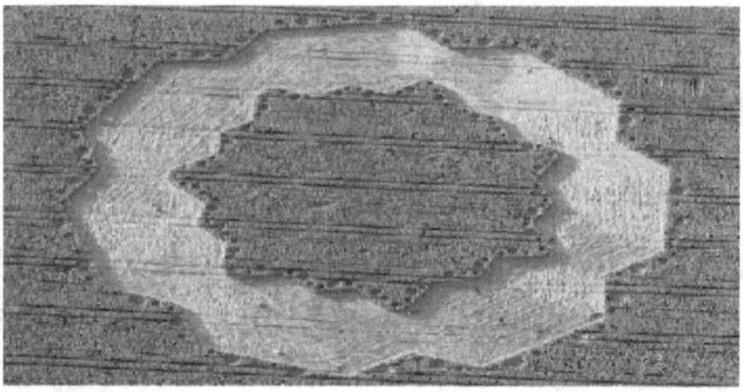

Figure 2-9: ***A Crop Circle Depicting Galactic Civilization. For details, see*** **book 《Crop Circle》.**

Chapter Overview:

In this chapter, we examine five representative structural domains of chemistry:

- **The Periodic Table** — the foundation of material organization;
- **Electronic Shell Structure** — microscopic structural constraints;
- **Bonding Regularities** — the generative rules of molecular formation;
- **Chemical Reaction Networks** — pathways of structural evolution;
- **Molecular Complexity Hierarchies** — large-scale material organization.

Together, these constitute five fundamental structural layers of chemistry and form a complete five-stage chain of structural generation. In summary, **structural periodicity in chemistry** may be expressed as: **Periodic Table → Electronic Shell Structure → Bonding Regularities → Chemical Reaction Networks → Molecular Complexity Hierarchies**

This progression corresponds closely to the structural frameworks previously established in mathematics and physics.

Section 1. Fundamental Prototypes of Structural Periodicity in Chemistry — Material Structures

Abstract:

The previous chapters examined structural periodicity in mathematics and physics. Mathematics revealed abstract prototypes of structural generation and stability, while physics demonstrated structural periodicity within natural dynamical systems.

In chemistry, structural periodicity appears through the generation and organization of material systems. Unlike mathematics and physics, chemistry deals directly with tangible material structures, and its periodicity is therefore manifested through stable material configurations and recurring patterns of chemical reactions.

From the perspective of the Structural Periodic Table, five representative structural prototypes can be identified:

- the Periodic Table,
- electronic shell structure,
- bonding regularities,
- chemical reaction networks,
- molecular complexity hierarchies.

Together, these prototypes constitute the periodic foundation of hierarchical material structures.

I. The Periodic Table: The Fundamental Periodicity of Material Structure

The most familiar manifestation of structural periodicity in chemistry is the Periodic Table.

When elements are arranged according to atomic number, numerous properties display recurring periodic patterns, including:

- atomic radius,
- electronegativity,
- chemical reactivity.

These regularities demonstrate that elements do not exist randomly, but are organized according to stable structural hierarchies.

The Periodic Table may therefore be interpreted as the manifestation of structural periodicity at the elemental level, revealing the fundamental organization of stable material structures.

II. Electronic Shell Structure: Microscopic Constraints on Stability

The origin of elemental periodicity lies in electronic shell structure. The arrangement of electrons within atoms follows quantum-mechanical principles, including:

- the principal quantum number,
- the orbital angular momentum quantum number,
- the spin quantum number.

These principles determine stable shell configurations such as:

2, 8, 18, 32, ...

Whenever an electronic shell becomes filled, the corresponding element exhibits exceptional stability, as exemplified by the noble gases. Electronic shell structure therefore provides the microscopic constraints responsible for elemental periodicity.

III. Bonding Regularities: The Generative Rules of Molecular Structure

When elements combine to form molecules, a new level of structural periodicity emerges.

Representative bonding mechanisms include:

- covalent bonds,
- ionic bonds,
- metallic bonds,
- hydrogen bonds.

These bonding patterns obey well-defined structural rules that enable molecules to form stable configurations, including:

- carbon frameworks in organic compounds,
- inorganic crystal lattices.

Within the framework of structural periodicity, chemical bonds represent the fundamental generative rules governing material structures.

Through these rules, relatively simple elements become capable of generating highly complex molecular architectures.

IV. Chemical Reaction Networks: Pathways of Structural Evolution

If chemical bonds govern structural generation, chemical reactions govern structural evolution.

In complex systems, reactions rarely occur as isolated events but instead form interconnected networks, such as:

- catalytic reaction networks,
- metabolic networks,
- organic reaction pathways.

These networks describe the transformations of matter among different structural states.

From the viewpoint of structural periodicity, reaction networks represent the evolutionary pathways of material structures through time. They determine which structures emerge, remain stable, or disappear.

V. Molecular Complexity Hierarchies: Macroscopic Structural Organization

As molecular size increases, structural complexity also increases.

Representative examples include:

- small molecules,
- polymers,
- biological macromolecules,
- complex chemical systems.

These hierarchical levels exhibit significant structural differences, including:

- organic molecular architectures,
- protein folding,
- self-assembled systems.

Such phenomena demonstrate that chemical systems possess well-defined hierarchical complexity.

From the perspective of structural periodicity, these hierarchies represent the macroscopic expression of periodic organization within material structures.

VI. Summary

Through the five structural prototypes —

- the Periodic Table,
- electronic shell structure,
- bonding regularities,
- chemical reaction networks,
- molecular complexity hierarchies —

we may identify the evolutionary pathway of structural periodicity in chemistry: **Elemental Periodicity → Shell Constraints → Bond Formation → Reaction Network Evolution → Molecular Complexity Hierarchies**

This sequence illustrates the hierarchical development of material structures from microscopic organization to macroscopic complexity.

Chemistry therefore not only reveals the composition of matter but also demonstrates how structural periodicity organizes material systems across multiple hierarchical levels.

In the following section, these structures will be reinterpreted within the ternary structural framework, showing how elements, reaction pathways, and molecular structures form a unified structural cycle.

By this point, the three scientific disciplines exhibit a remarkably symmetric structural organization.

Discipline	**Five Structural Prototypes**
Mathematics	Recursion • Groups • Periodic Functions • Moduli Spaces • Structural Spectra
Physics	Particle Spectra • Field Modes • Conservation Laws • Phase Transitions • Cosmic Hierarchies
Chemistry	Periodic Table • Electronic Shell Structures • Bonding Regularities • Chemical Reaction Networks • Molecular Complexity Hierarchies

Table 2-9-1-1: Five Structural Prototypes in Mathematics, Physics, and Chemistry.

Together, these three parallel frameworks establish a coherent and highly symmetric theoretical foundation for the Structural Periodic Table.

Section 2. A Ternary Interpretation of Structural Periodicity in Chemistry

Abstract:

In the previous section, we examined five fundamental prototypes of structural periodicity in chemistry: the Periodic Table, electronic shell structure, bonding regularities, chemical reaction networks, and molecular complexity hierarchies. Together, these structures form the hierarchical organization of material systems, from the microscopic to the macroscopic scale.

From a deeper structural perspective, these diverse chemical structures can be understood as manifestations of the interaction among three fundamental structural dynamics:

- **Point (•):** the basic units of material structure;
- **Line (|):** chemical reactions and pathways of structural evolution;
- **Circle (O):** the stable closure of molecular and material structures.

These three structural elements are not merely geometric symbols, but abstract representations of the three fundamental mechanisms governing **generation, evolution, and stability** in chemical systems.

I. Point (•): Elements as the Fundamental Units of Material Structure

Within chemical systems, all complex structures ultimately originate from the chemical elements.

Elements consist of atoms, whose electronic shell structures determine their chemical properties. Differences in electronic configurations give rise to the periodic patterns observed in the Periodic Table.

From the perspective of structural periodicity, elements may therefore be regarded as the fundamental generative units of material structures. These units provide the starting point for the formation of material organization. For example:

- hydrogen and oxygen combine to form water molecules;
- carbon forms an enormous variety of organic structures.

The combinatorial possibilities among different elements provide the fundamental building blocks of the material world.

II. Line (|): Chemical Reactions as Pathways of Structural Evolution

If elements provide the generative units, chemical reactions provide the pathways of structural evolution.

Within chemical systems, material structures continuously transform through reactions such as:

- oxidation–reduction reactions;
- acid–base reactions;
- organic synthetic pathways.

In complex systems, these reactions frequently form interconnected networks, including:

- metabolic networks;
- catalytic cycles;
- organic reaction maps.

These networks describe the transformations among different material structures.

From the viewpoint of structural periodicity, reaction pathways represent the evolutionary trajectories of material structures through time. They connect different structural states and drive systems toward new stable configurations.

III. Circle (O): Molecular Structures and Stable Closure

When elements combine through chemical reactions, stable structural closures emerge, including:

- molecular structures;
- crystal structures;
- polymeric structures.

These structures often exhibit characteristic stable geometries, such as:

- cyclic molecules;
- crystal lattices;
- folded protein structures.

A stable structure signifies that the interactions among atoms have reached equilibrium.

Within the language of structural periodicity, this corresponds to the formation of structural closure under appropriate constraints.

Such closed structures provide the stable material forms upon which chemical systems depend.

IV. The Cyclic Relationship of the Ternary Structure

Within chemistry, the three structural dynamics — Point, Line, and Circle — also form a continuous cycle: **Elements (Point) → Reaction Pathways (Line) → Molecular Structures (Circle) → New Combinations of Elements (New Point)**

This cycle explains the generation and evolution of material structures:

1) Elements provide the fundamental structural units.
2) Chemical reactions drive structural transformation.
3) Molecular structures achieve stable closure.
4) Newly formed structures become the basis for subsequent reactions.

For example:

• simple elements combine to form molecules;

• molecules participate in further reactions to generate increasingly complex structures;

• these new structures subsequently enter additional chemical processes.

This recurring cycle constitutes one of the fundamental mechanisms underlying the continual evolution of the material world.

V. Hierarchical Manifestations of the Ternary Structure in Chemistry

Within chemical systems, the ternary structure appears at different hierarchical levels.

Structural Element	Chemical Manifestation
Point (•)	Elements and atoms
Line (|)	Chemical reaction pathways
Circle (O)	Molecular and crystal structures

Table 2-9-2-1: Hierarchical manifestations of the ternary structure in chemistry.

This relationship suggests that complex chemical structures are not formed randomly, but instead follow a cyclic process of generation, evolution, and stabilization.

Accordingly, profound structural relationships exist among elemental periodicity, reaction networks, and molecular structures.

VI. Summary

From the perspective of structural periodicity, material structures in chemistry may be interpreted through the three fundamental structural dynamics:

• **Point (•):** elements and atoms as structural origins;

• **Line (|):** chemical reaction pathways as evolutionary processes;

• **Circle (O):** molecular structures as stable structural closure.

Together they form the recurring cycle: **Generation → Evolution → Stability → New Structural Generation**

This cycle explains not only the formation of elemental periodicity and molecular structures, but also the hierarchical evolution of material systems from simple elements to highly complex molecules.

In the next section, we shall examine how the Structural Periodic Table can be applied within chemistry by analyzing chemical reaction networks, molecular configurations, and elemental periodicity as concrete and testable examples of structural periodicity.

Overall Summary: By this point, the three scientific disciplines have formed a remarkably elegant **ternary structural symmetry**.

Discipline	Point (•)	Line (|)	Circle (O)
Mathematics	Initial Rules	Recursive Development	Structural Closure
Physics	Local Sources	Dynamical Propagation	Stable Structures
Chemistry	Elements	Reaction Pathways	Molecular Structures

Table 2-9-2-2: The ternary symmetry of Point, Line, and Circle in mathematics, physics, and chemistry.

This table captures one of the central ideas of the entire book: Point (•), Line (|), and Circle (O) constitute the three fundamental structural drivers underlying the generation, evolution, and stabilization of complex systems.

Section 3. From Chemical Structural Periodicity to Structural Periodicity in the Natural Sciences

This chapter has examined structural periodicity from the perspective of chemistry, focusing on its fundamental manifestations within the hierarchical organization of material systems.

Through five representative structural prototypes — the **Periodic Table**, **electronic shell structure**, **bonding regularities**, **chemical reaction networks**, and **molecular complexity hierarchies** — we have identified the characteristic evolutionary pathway of material structures from the microscopic to the macroscopic scale: **Elemental Periodicity → Shell Constraints → Bond Formation → Reaction Network Evolution → Molecular Complexity Hierarchies**

This progression demonstrates that chemical structures do not arise randomly, but gradually organize into hierarchical systems through the interplay between structural constraints and chemical evolution.

Furthermore, this chapter has provided a unified interpretation of these phenomena through the ternary structural language. Within this framework:

- **Point (•)** represents the fundamental units of material structure, such as elements and atoms;
- **Line (|)** represents evolutionary pathways through time, such as chemical reactions and reaction networks;
- **Circle (O)** represents stable structural closure under physical and chemical constraints, such as molecular and crystalline structures.

Together they form the recurring cycle: **Point → Line → Circle → New Point**

This cycle reveals the fundamental mechanism governing the generation and evolution of material structures. Elements serve as structural origins; reaction pathways generate new configurations; and stable molecular or crystalline structures become the starting point for subsequent cycles of structural development.

From this perspective, elemental periodicity, chemical bonding, and molecular complexity are not isolated chemical phenomena, but rather different manifestations of the same ternary structural cycle operating within material systems.

Comparison with the previous chapters reveals a clear hierarchical correspondence across the natural sciences.

Discipline	Point (Generation)	Line (Evolution)	Circle (Stability)
Mathematics	Initial Rules	Recursive Development	Structural Closure
Physics	Local Sources	Dynamical Propagation	Stable Modes
Chemistry	Elements	Reaction Pathways	Molecular Structures

Table 2-9-3-1: Structural Periodicity Across the Natural Sciences.

This correspondence indicates that, although mathematics, physics, and chemistry investigate different domains, they share the same underlying structural cycle of **Generation → Evolution → Stability**

Consequently, structural periodicity in chemistry is not merely an organizing principle of the material world, but also an essential component of structural periodicity throughout the natural sciences: **Chemistry → Ternary Structure → Natural Sciences**

Through this hierarchical progression, the present volume establishes a continuous structural framework across the natural sciences: **Mathematical Structural Periodicity → Physical Structural Periodicity → Chemical Structural Periodicity**

This framework suggests that complex structures in nature do not emerge randomly, but arise through recurring cycles of generation, evolution, and stabilization.

At higher levels of complexity — such as biological and social systems — similar structural cycles may continue to operate. Their periodicity, however, is likely to appear less as strict mathematical regularity and more as statistical or structural tendencies.

Accordingly, chemical structural periodicity may be regarded both as the culmination of structural periodicity within the natural sciences and as an important bridge toward the study of more complex systems.

Having established the logical, dynamical, and material foundations of structural periodicity, we now proceed to examine how these same structural principles extend into living systems, where organization, adaptation, and evolution introduce new layers of complexity.

Concluding Note: Mathematics reveals the logical foundations of structural periodicity. Physics demonstrates its dynamical realization. Chemistry embodies its material manifestation. Together, these three disciplines establish the foundational pillars of the Structural Periodic Table.

References (Chapter Nine)

The following references provide background for chemical periodicity, electronic shell structure, molecular bonding, reaction networks, and molecular complexity. They situate the structural interpretation of chemical systems within established chemical theory.

[1] Mendeleev, D. (1869). The Relation Between the Properties and Atomic Weights of the Elements. *Journal of the Russian Chemical Society*.

[2] Atkins, P. & Paula, J. de (2018). *Physical Chemistry* (11th ed.). Oxford University Press.

[3] Levine, I. N. (2014). *Quantum Chemistry* (7th ed.). Pearson.

[4] Pauling, L. (1960). *The Nature of the Chemical Bond* (3rd ed.). Cornell University Press.

[5] Barabási, A. L. (2016). *Network Science*. Cambridge University Press.

[6] Whitesides, G. M. & Grzybowski, B. (2002). Self-Assembly at All Scales. *Science*, 295(5564), 2418–2421.

[7] Ertl, G. (2008). Reactions at Solid Surfaces. *Nobel Lecture*. Angewandte Chemie International Edition.

[8] Laughlin R. B. & Pines, D. (2000). The Theory of Everything. *Proceedings of the National Academy of Sciences*, 97(1), 28–31.

[9] Anderson, P. W. (1972). More Is Different. *Science*, 177(4047), 393–396.

[10] Chang, J. (2026). *Recursive Self-Organizing Evolutionary Breakthrough — Unifying Natural, Social and Life Sciences.* Universal Publishing.

Chapter Ten: Chemical Structural Periodicity and the Origin of Life

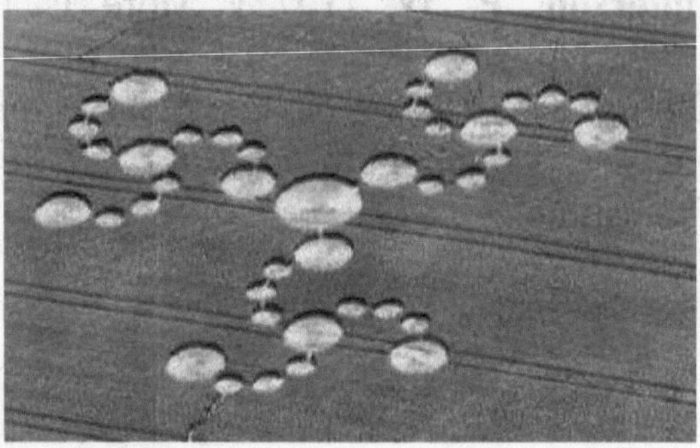

Figure 2-10: A crop circle illustrating a Galactic Civilization. For details, see《*Crop Circle*》.

Chapter Overview:

The preceding chapters have examined the manifestations of the Structural Periodic Table in mathematics, physics, and chemistry. In this framework:

- **Mathematics** investigates the language of structure.
- **Physics** investigates the dynamics of nature.
- **Chemistry** investigates the organization of matter.

Yet a deeper question naturally arises: **when elements, chemical reactions, and molecular architectures become sufficiently complex, why does life emerge?**

This question forms the central theme of the present chapter.

Traditional studies on the origin of life usually approach this problem from several perspectives, including:

- elemental composition,
- chemical reactions,
- the RNA world hypothesis,
- cellular formation,

- and biological evolution.

This chapter adopts the perspective of the **Structural Periodic Table**, integrating these different viewpoints into a unified structural framework organized into four successive stages.

Section	Central Question	Structural Role
1	Why does life select particular elements?	Point •
2	Why does a dynamic chain of life emerge?	Line |
3	Why can life form stable structures?	Circle O
4	When does life truly begin?	Triadic Structural Closure

Table 2-10-1: Four-stage structural framework for the origin of life.

The first three sections correspond to three fundamental levels involved in the emergence of life.

Structural Level	Biological Stage
Point •	Life window (CHNOPS elements)
Line |	Dynamic life chains (autocatalysis, metabolism, RNA cycles)
Circle O	Structural closure (DNA, cell membranes, protein architectures)

Table 2-10-2: Three structural levels underlying the emergence of life.

Once these three levels become integrated into a unified system, the discussion naturally proceeds to the central question addressed in **Section 4**: **Where is the true boundary between chemistry and life — from viruses to bacteria?**

Consequently, this chapter establishes a continuous structural sequence: **Elements → Reactions → Configurations → Living Systems**, corresponding to **Point • → Line ｜ → Circle O → Life.**

From the perspective of the Structural Periodic Table, life is not regarded as an abrupt event, but as the natural consequence of the progressive closure of **matter, dynamics, and structure**. In this sense, the discussion extends beyond traditional chemistry and enters the transitional domain between chemistry and the life sciences.

At the same time, the three fundamental structural levels discussed throughout this chapter correspond to the three principal validation directions of the Structural Periodic Table.

Validation Direction	**Structural Correspondence**
Elemental periodicity	Electronic shells and material structures
Reaction networks	Dynamic structures
Molecular configurations	Geometric structures

Figure 2-10-3: **Three validation levels for the origin of life.**

If these three levels are indeed governed by a common structural principle, then the origin of life may represent not only a biological problem, but also a problem of **structural periodicity**.

Therefore, the central question of Chapter 10 is not simply **how life evolved**, but rather: **Why can life emerge naturally from chemical structure?**

In summary, **chemistry validates whether structure possesses material realizability.**

Section 1. The Periodic Table and the Material Foundation of Life – From the Elemental Window to the Life Window

Abstract:

The emergence of life depends on a specific combination of chemical elements. Modern biochemistry shows that the core elements of living systems are primarily hydrogen (H), carbon (C), nitrogen (N), oxygen (O), phosphorus (P), and sulfur (S). Although more than one hundred chemical elements are known to exist in nature, only a small subset constitutes the principal material basis of life.

Traditional chemistry explains this phenomenon mainly through atomic structure and chemical properties. From the perspective of the **Structural Periodic Table**, this chapter proposes a different interpretation: the selection of life's building blocks is not entirely accidental, but is associated with a special **Life Window** within the periodic table. This window is jointly determined by elemental stability, bonding capability, and structural generative potential, thereby providing the material foundation necessary for the emergence of life.

I. Why Does Life Select Only a Small Number of Elements?

Observation of terrestrial life reveals that nearly all living systems are primarily built from the following elements: **Hydrogen (H); Carbon (C); Nitrogen (N); Oxygen (O); Phosphorus (P); Sulfur (S).**

Collectively known as the **CHNOPS system**, these elements constitute the fundamental components of: **water, proteins, DNA, RNA and cell membranes.**

However, the periodic table contains well over one hundred known elements.

This naturally raises an important question: **Why is life not based upon gold (Au), lead (Pb), or uranium (U)?**

II. The Conventional Chemical Explanation

Modern chemistry generally explains the selection of biological elements by emphasizing three major characteristics.

1) Moderate Atomic Mass

These elements possess suitable atomic sizes and electronic structures that enable the formation of stable chemical bonds.

2) Strong Bonding Capacity

Carbon, in particular, can form four covalent bonds simultaneously, allowing the construction of extraordinarily complex molecular networks.

3) High Cosmic Abundance

Hydrogen, oxygen, carbon, and several other life elements are among the most abundant elements in the universe, making them readily available during planetary evolution.

These explanations are undoubtedly important.

However, they primarily describe **what properties these elements possess**, rather than explaining **why these favorable properties are concentrated within only a small group of elements.**

III. The Life Window in the Structural Periodic Table

From the perspective of the Structural Periodic Table, the elements supporting life are not randomly distributed throughout the periodic table. Instead, they occupy a distinctive **structural window**, referred to here as the **Life Window**.

This window simultaneously satisfies three fundamental structural requirements:

1) Stability — the elements themselves remain sufficiently stable over long periods.

2) Connectivity — they readily form extensive structural linkages through diverse chemical bonds.

3) Evolvability — they continually generate new structural combinations and increasing molecular complexity.

If any one of these three conditions is absent, the emergence of life becomes highly improbable.

Consequently, life does not arise from all chemical elements equally, but rather from those lying within a structural window satisfying these three essential conditions.

IV. The Point • Structure as the Material Source of Life

Within the Structural Periodic Table, the **Point (•)** represents the **origin of structural generation**.

At the earliest stage of life's emergence, neither complex ecosystems nor fully developed cells were required. Instead, life first required **a reservoir of combinable materials**.

Accordingly, the CHNOPS elements may be interpreted as representing the **Point layer** of biological structure.

They constitute the primordial material library from which life could eventually emerge.

V. Why Are Gold, Lead, and Uranium Unsuitable as the Basis of Life?

From a structural perspective:

- **Gold (Au)** is chemically stable but relatively inert, making it unsuitable for constructing highly interconnected molecular networks.
- **Lead (Pb)** can participate in chemical reactions but exhibits comparatively limited structural diversity.
- **Uranium (U)** contains abundant nuclear energy but lacks long-term structural stability; its radioactive decay continuously disrupts complex molecular organization.

Consequently, while such elements may participate in the environment surrounding life, they are unlikely to become the principal structural materials of living systems.

VI. The Life Window and the Periodic Table

From the perspective of the Structural Periodic Table, the periodic table is more than a chemical classification of elements.

It is also a **map of structural possibilities**.

Each element represents a different structural capability, while the essential elements of life are concentrated within the region most favorable for generating complex and evolvable structures.

Accordingly, the **Life Window** is not merely the result of biological selection; rather, it reflects a long-term structural selection embedded within the periodic organization of the elements themselves.

VII. The First Step from Elements to Life

The first stage in the origin of life is not life itself, but the establishment of suitable material conditions.

Only when elements simultaneously satisfy the three structural requirements — **"stability → connectivity → evolvability"** — can biological organization begin to emerge.

Thus, the periodic table determines not only chemical reactions but also the structural boundary of biological possibility.

VII. Conclusion

Life does not arise randomly from all chemical elements. Instead, the principal building blocks of living systems are concentrated within a distinctive structural region—the **Life Window**. This window simultaneously possesses:

- structural stability,
- extensive bonding capability,

• and continuous evolutionary potential.

Together these properties provide the material foundation required for life's emergence. Within the Structural Periodic Table, these elements correspond to the **Point (•) layer** of living systems — the structural source that precedes biological organization.

Therefore, the first step in the origin of life is not the appearance of the first cell, but the formation of the **Life Window** within the periodic table itself.

Closing Remark: *The Periodic Table determines which atoms are capable of existing; the Life Window determines which of those atoms are capable of becoming life.*

Note: The complete list of references appears at the end of this chapter. Two works are particularly relevant to this section:

- **Schrödinger, E. (1944).** ***What Is Life?***
- **Kauffman, S. A. (1993).** ***The Origins of Order: Self-Organization and Selection in Evolution.***

Both works address a fundamental question: **Why does life develop organized structure?**

This question closely parallels the central concern of the **Structural Periodic Table.**

The concept of the **Life Window** proposed in this section extends these earlier perspectives one step further. Schrödinger emphasized the physical conditions necessary for life, while Kauffman explored how life may arise through self-organization. The present work instead asks **why only a particular structural window within the periodic table possesses the capacity to serve as the material source of life.**

This interpretation corresponds directly to the **Point (•) layer** of the Structural Periodic Table, representing the fundamental material origin of biological structure.

Section 2. Reaction Networks and the Dynamical Origin of Life – From Chemical Reactions to the Life Dynamic Chain

Abstract:

Research on the origin of life has long sought to explain how living systems first emerged. However, the existence of biological building blocks alone is insufficient to produce life. Numerous experiments have demonstrated that many organic molecules can form under natural conditions, yet a far more fundamental question remains: **why are these molecules able to organize themselves into persistent and stable systems?**

From the perspective of the **Structural Periodic Table**, this section proposes that the key to life lies not in individual molecules, but in **reaction networks capable of maintaining themselves over time**. Autocatalytic reactions, metabolic networks, and RNA cycles all exhibit a common characteristic: reactions promote further reactions, thereby forming a continuously evolving **Life Dynamic Chain**. Within the structural language developed in this book, this stage corresponds to the **Line (|) layer**, representing the critical transition from static chemical materials to dynamic living systems.

I. Why Is Life More Than a Collection of Molecules?

Modern biology shows that living organisms are primarily composed of: proteins, **nucleic acids, lipids, and carbohydrates.**

Yet even if all these molecules are artificially mixed together, life does not spontaneously emerge.

Therefore, a fundamental distinction exists between living systems and ordinary chemical mixtures.

The essential question is not: **What molecules are present?**

Rather, it is: **How do these molecules continuously interact with one another?**

II. What Does Life Truly Require?

From the viewpoint of the Structural Periodic Table, the emergence of life requires two fundamental conditions.

1) The Life Window

Suitable biological materials must first become available. This corresponds to the **Life Window**, discussed in the previous section.

2) The Life Dynamic Chain

Chemical reactions must become self-sustaining.

Only when reactions begin forming interconnected networks does the possibility of life arise. Accordingly, the essence of life is not simply a collection of molecules, but rather a **self-sustaining reaction network**.

III. Autocatalysis: The Earliest Form of Biological Dynamics

One of the most important concepts in origin-of-life research is **autocatalysis**. In an autocatalytic system, the products of a reaction promote the continuation of the same reaction. For example,

$$A + B \rightarrow C$$

while the newly produced **C** accelerates additional reactions, producing even more **C**. This establishes a positive feedback cycle:

$$A + B \rightarrow C \rightarrow \{\text{more}\}\, C$$

Such systems possess an inherent capacity for **self-amplification**.

From the perspective of the Structural Periodic Table, this represents the simplest manifestation of the **Line (|) layer**, since the structure has begun to drive its own evolution.

IV. Metabolic Networks: The Formation of the Life Dynamic Chain

Modern life depends upon highly organized metabolic systems, including:

- glycolysis,
- the citric acid (Krebs) cycle,
- and the respiratory chain.

Their common characteristic is that the product of one reaction becomes the substrate for the next.

Thus a continuous reaction sequence is established:

$$\mathbf{R_1 \rightarrow R_2 \rightarrow R_3 \rightarrow R_4}$$

These interconnected pathways not only generate energy but also maintain structural organization.

Consequently, metabolic networks may be regarded as the mature expression of the **Life Dynamic Chain**.

V. The RNA Cycle: The Union of Dynamics and Information

Among the major hypotheses concerning the origin of life, the **RNA World Hypothesis** occupies a central position.

RNA possesses two remarkable capabilities:

- storing genetic information,
- catalyzing chemical reactions.

Under suitable conditions, RNA molecules can even facilitate their own replication. This gives rise to a continuous cycle:

RNA → Replication → More RNA

Within the Structural Periodic Table, the significance of RNA lies not merely in its chemical composition, but in its ability to integrate **dynamic processes with information storage**.

This represents a decisive step toward biological organization.

VI. From Individual Reactions to Reaction Networks

Ordinary chemical reactions generally proceed as:

$A+B \rightarrow C$

After completion, the system approaches chemical equilibrium.
Living systems behave fundamentally differently. Instead of isolated reactions, they establish cyclic reaction networks:

$$\mathbf{R_1 \rightarrow R_2 \rightarrow R_3 \rightarrow R_1}$$

In such systems, reactions cease to be isolated events and become continuous processes. Therefore, **life is not a single chemical reaction; it is a reaction network.**

VII. The Line (|) Structure and the Life Dynamic Chain

Within the Structural Periodic Table, the **Line (|)** represents:

- evolution,
- propagation,
- and continuous structural generation.

The Life Dynamic Chain displays precisely these characteristics. Specifically,

- autocatalysis drives amplification,
- metabolic networks sustain circulation,
- RNA systems enable replication.

Together these mechanisms demonstrate that biological structure has begun to propagate continuously through time.

Accordingly, the second stage in the origin of life corresponds to the **Line (|) layer**, namely the emergence of the **Life Dynamic Chain**.

VIII. The Second Step from Chemistry to Life

The emergence of life was not an instantaneous event. From a structural perspective, it required at least two successive stages.

Stage One: Formation of the **Life Window**, providing the necessary chemical materials.

Stage Two: Formation of the **Life Dynamic Chain**, allowing reaction networks to become self-sustaining.

Only when both the **material foundation** and the **dynamic foundation** coexist can a system approach the threshold of genuine life.

IX. Conclusion

The defining characteristic of life is not molecular complexity itself, but the existence of **reaction networks capable of sustaining and reproducing their own organization**.

Although autocatalytic systems, metabolic networks, and RNA cycles differ chemically, they all exhibit the same structural principle: **structure continuously generates further structure.**

Therefore, the second step in the origin of life is not the production of new molecules, but the emergence of a new **Life Dynamic Chain**.

Within the Structural Periodic Table, this stage corresponds to the **Line (|) layer**, representing the earliest evolutionary mechanism of living systems.

Closing Remark: *Elements make life possible; reaction networks set life in motion. When structure becomes capable of continuously generating new structure, life takes its second decisive step toward existence.*

Note: The complete reference list is provided at the end of this chapter. The following four works are particularly relevant to this section because they closely correspond to its three central structural concepts.

Structural Concept	Representative Reference
Autocatalytic reactions	Kauffman (1986)
Metabolic networks	Morowitz (1992)
RNA cycles	Gilbert (1986)
Self-organizing life	Schrödinger (1944)

Figure 2-10-2-1. Correspondence between the Structural Periodic Table and four classic works on the origin of life.

Ultimately, this section arrives at the following structural interpretation:

Life ≠ A collection of molecules
Life = A self-sustaining reaction network

This conclusion is closely aligned with the central ideas of:

- **Erwin Schrödinger's *What Is Life?***
- **Stuart Kauffman's *The Origins of Order***

The principal contribution of the present work is to place these ideas within the unified structural framework of: **Point (•) → Line (|) → Circle (O)**

and to identify this stage explicitly as the **Line (|) layer: the Life Dynamic Chain**. This concept serves as the dynamic bridge between the **Life Window** introduced in Section 1 and the **Life Closure** to be discussed in the following section.

Section 3. Molecular Configurations and the Structural Closure of Life – From Chemical Architecture to the Formation of Living Systems

Abstract:

Life requires not only appropriate chemical elements but also continuously operating reaction networks. Nevertheless, neither suitable materials nor dynamic processes alone are sufficient to generate life. The defining characteristic that distinguishes living systems from ordinary chemical systems is their ability to maintain their own organization over extended periods.

Modern biology demonstrates that DNA double helices, cellular membranes, and protein folding all exhibit highly stable three-dimensional configurations. From the perspective of the **Structural Periodic Table**, this chapter proposes that the third decisive step in the origin of life is not the emergence of new reactions, but the achievement of **structural closure**. By establishing stable molecular architectures, living systems preserve matter, energy, and information simultaneously, thereby acquiring the capacity for long-term self-maintenance. Within the Structural Periodic Table, this stage corresponds to the **Circle (O) layer**, representing the **Life Structural Closure**.

I. Why Does Life Not Simply Fall Apart?

From a chemical perspective, living organisms consist of enormous numbers of molecules.

Yet most ordinary chemical systems eventually:

- decompose,
- become diluted,
- or approach thermodynamic equilibrium,

thereby losing their structural organization.

This raises a fundamental question: **Why are living systems capable of maintaining themselves for long periods of time?**

If life were merely an accumulation of complex molecules, it should rapidly disintegrate. Reality clearly shows otherwise.

This suggests that living systems possess a unique mechanism for structural stability.

II. What Does Life Ultimately Require?

The previous two sections introduced the first two stages in the emergence of life.

Stage One: the **Life Window**, providing the necessary material foundation.

Stage Two: the **Life Dynamic Chain**, establishing self-sustaining reaction networks.

However, even continuously operating reactions cannot produce life unless the system possesses a stable organizational boundary.

Consequently, life requires a third essential condition: **structural closure.** Only when a system becomes capable of maintaining its own organization does genuine life emerge.

III. Why Is the DNA Double Helix Stable?

DNA serves as the principal carrier of genetic information.

Its defining characteristic is its **double-helical structure**, which provides several important advantages:

- complementary base pairing,
- stabilization through hydrogen bonding,
- faithful storage and replication of genetic information.

More fundamentally, the double helix forms a **closed information architecture**.

From the perspective of the Structural Periodic Table, the significance of DNA lies not merely in its chemical composition, but in its capacity for the long-term preservation of information.

Accordingly, DNA represents one of the most important examples of **Life Structural Closure**.

IV. Why Is the Cell Membrane Stable?

Living systems must preserve not only information but also their own physical boundaries. The cell membrane, primarily composed of a phospholipid bilayer, provides this capability through:

- separation of internal and external environments,
- selective transport,
- maintenance of energetic balance.

As a consequence, a stable boundary emerges between the organism and its surroundings. From a structural perspective, the cell membrane accomplishes **system closure**.

For the first time, life acquires a distinction between **self** and **non-self**. Therefore, the cell membrane constitutes the physical boundary of biological structural closure.

V. Why Is Protein Folding Stable?

Proteins are synthesized as linear chains of amino acids.

Their biological functions, however, arise not from the linear sequence itself but from their **three-dimensional folded configurations**. Protein folding generates:

- active sites,
- catalytic centers,
- functional molecular architectures.

Only after reaching a stable conformation can proteins perform their biological roles.

Thus, protein folding illustrates how **structural function emerges through spatial closure**.

VI. From Molecules to Living Systems

Although DNA, cell membranes, and proteins perform different biological functions, they share one essential characteristic: **they establish stable structural configurations.**

They are no longer merely products of chemical reactions. Instead, they become structural units capable of maintaining themselves. Consequently, the defining characteristic of life is not molecular abundance, but **structural closure**.

VII. The Circle (O) Structure and Life Structural Closure

Within the Structural Periodic Table, the **Circle (O)** represents:

- stability,
- closure,
- self-maintenance.

Living systems exhibit precisely these characteristics.

- DNA preserves information.
- Cell membranes preserve organizational boundaries.
- Proteins preserve biological functions.

Together they establish a unified **Life Structural Closure**.

Only at this stage does a biological system acquire the capacity for long-term persistence.

VIII. The Third Step from Chemistry to Life

From a structural perspective, the emergence of life proceeds through three successive stages.

Stage One: Life Window — the emergence of suitable elemental building blocks.
Stage Two: Life Dynamic Chain — the establishment of self-sustaining reaction networks.

Stage Three: Life Structural Closure — the formation of stable molecular architectures.

Only when these three conditions coexist does life truly emerge.

Accordingly, the defining event in the origin of life is not the appearance of a particular molecule, but the achievement of **structural closure**.

IX. Conclusion

Although DNA double helices, cell membranes, and protein folding belong to different levels of biological organization, they all illustrate the same structural principle: **stable configurations enable long-term existence.**

Therefore, the third step in the origin of life is not the creation of new chemical reactions, but the emergence of **Life Structural Closure**.

Within the Structural Periodic Table, this corresponds to the **Circle (O) layer**, through which living systems acquire the capacity for self-maintenance.

X. The Structural Transition from Chemistry to Life

Combining the first three sections yields a unified structural pathway for the emergence of life.

Step One: Elemental Window → Life Window (Material foundation)

Step Two: Reaction Network → Life Dynamic Chain (Dynamic foundation)

Step Three: Molecular Configuration → Life Structural Closure (Structural foundation)

Together they establish the complete progression: **Elements → Reactions → Configurations → Life,**

corresponding structurally to **Point (•) → Line (|) → Circle (O) → Living Systems**

Life therefore does not appear suddenly. Instead, it emerges naturally as structural closure is progressively achieved.

Closing Remark: *Elements provide the materials of life. Reaction networks provide its dynamics. Molecular configurations provide its stability. When materials, dynamics, and structural closure coexist, chemistry crosses its boundary and becomes life.*

XI. Summary

Taken together, the first three sections establish the following correspondence between **Structural Periodicity** and the **Origin of Life.**

Structural Periodicity	**Origin of Life**
Point (•)	Life Window (CHNOPS)
Line (∣)	Life Dynamic Chain (Autocatalysis, Metabolism, RNA)
Circle (O)	Life Structural Closure (DNA, Membranes, Proteins)

Figure 2-10-3-1: Structural Periodicity and the Origin of Life.

The complete structural progression becomes: **{Elements} → {Reactions} → {Configurations} → {Life}**

This represents an upward transition from the framework established in the Chemistry Volume: **{Elements} → {Reactions} → {Compounds}**

to the broader question addressed by the **Structural Periodic Table**: **How does chemistry cross the boundary to become life?**

Note: The complete reference list appears at the end of this chapter. One work is particularly significant for this section:

Maturana, H. R., & Varela, F. J. (1980). ***Autopoiesis and Cognition: The Realization of the Living.***

Maturana and Varela introduced the influential concept of **autopoiesis**, proposing that the defining characteristic of life is neither DNA nor proteins themselves, but the capacity of a system to **continuously maintain its own organization and boundaries**.

This idea closely parallels the conclusion reached in this section:

- DNA preserves information.
- Cell membranes preserve boundaries.
- Proteins preserve biological function.

Together they establish **Life Structural Closure**.

Consequently, future development of the **Structural Periodic Table** may establish a meaningful theoretical connection between the **Circle (O)** as **Structural Closure** and the concept of **autopoiesis** in modern systems biology.

Such a connection would allow the **Circle (O)** to become more than a philosophical symbol, serving instead as a structural concept directly linked to contemporary theories of living systems.

Section 4. The Boundary of Life and the Simplest Living Systems – From Viruses to Bacteria

Abstract:

One of the central questions in origin-of-life research concerns the boundary between living and non-living systems. Modern biology has revealed that viruses and bacteria occupy opposite sides of this boundary. Viruses possess genetic information and structural stability, yet lack independent metabolic activity. Bacteria, by contrast, possess genetic information, metabolic networks, cellular boundaries, and autonomous replication.

From the perspective of the **Structural Periodic Table**, viruses and bacteria should not merely be regarded as two classes of microorganisms. Rather, they represent two critical stages in the emergence of life. Viruses embody the earliest appearance of structural closure, whereas bacteria represent the first complete living system in which material foundation, dynamic processes, and structural closure are fully integrated. Consequently, the transition from viruses to bacteria marks the crucial point at which chemical systems cross the boundary into biological life.

I. The Problem of the Boundary of Life

When does life truly begin?

This has long been one of the most fundamental questions in biology. For rocks, minerals, or ordinary chemical reactions, distinguishing living from non-living systems is straightforward.

Viruses, however, occupy a far more ambiguous position. Viruses possess:

- genetic material,
- protective structural shells,
- and the ability to replicate.

Nevertheless, they cannot perform independent metabolism.

Consequently, viruses have long been regarded as existing at the boundary between life and non-life.

II. Viruses: The Earliest Form of Life Structural Closure

A typical virus consists primarily of:

- DNA or RNA,
- a protein capsid,

and, in some cases, a lipid envelope. Structurally, viruses already exhibit several characteristics associated with living systems.

They can:

- store genetic information,
- undergo mutation and evolution,
- exploit host cells for replication.

However, they still lack several fundamental biological functions. Viruses cannot:

- perform independent metabolism,
- obtain energy autonomously,
- maintain internal homeostasis,
- reproduce outside a host organism.

Accordingly, viruses are more appropriately interpreted as **Boundary Structures**.

They have progressed beyond ordinary chemical molecules, yet have not achieved the status of complete living systems.

Within the Structural Periodic Table, viruses possess the capacities for **generation** and **propagation**, but their structural closure remains dependent upon the host cell. Consequently, they have not yet become fully autonomous living systems.

In structural terms, viruses approach the **Circle (O) layer**, while the **Line (|) layer** remains incomplete. They may therefore be regarded as **Proto-Life Systems**.

III. Bacteria: The First Complete Living System

Compared with viruses, bacteria possess:

- a DNA-based genetic system,
- a cellular membrane,
- metabolic networks,
- and autonomous reproduction.

Unlike viruses, bacteria no longer depend upon another living organism to sustain their existence. From a structural perspective, bacteria represent the first successful integration of:

- **Point (•):** biological material source,
- **Line (|):** Life Dynamic Chain,
- **Circle (O):** Life Structural Closure.

These three structural components become fully unified.

Accordingly, bacteria may be regarded as **the first complete living system**.

IV. The Structural Transition from Viruses to Bacteria

Traditional biology primarily emphasizes the taxonomic differences between viruses and bacteria.

The Structural Periodic Table instead focuses on the structural stages they represent. From this viewpoint:

- viruses possess information and structural organization,
- bacteria possess information, structure, and autonomous dynamics simultaneously.

Therefore, the essential distinction between viruses and bacteria is neither size nor complexity.

Rather, it is the capacity for **autonomous self-maintenance**.

V. Why Does Life Truly Begin with Bacteria?

If life is defined as **a system capable of maintaining its own existence over time**, then three conditions become indispensable.

1) Preservation of genetic information.
2) Continuous energy flow through metabolism.
3) Maintenance of a stable structural boundary.

Viruses satisfy only part of these requirements. Bacteria satisfy all three. Consequently, genuine life begins only when a complete structural loop is established.

VI. The Boundary of Life in the Structural Periodic Table

Combining the previous discussions yields four successive stages in the emergence of life.

Stage One: Life Window — elemental foundation.
Stage Two: Life Dynamic Chain — reaction networks.
Stage Three: Life Structural Closure — stable molecular architectures.
Stage Four: The Simplest Living System — the transition from viruses to bacteria.

Together they establish the complete structural progression: **Elements → Reactions → Configurations → Life**

VII. From Chemical Systems to Living Systems

Chemical systems are capable of:

- forming complex molecules,
- constructing reaction networks,
- establishing stable molecular architectures.

However, genuine life appears only when these structures become capable of:

- preserving information,
- acquiring energy,
- maintaining boundaries,
- reproducing themselves.

Life, therefore, should not be defined by the presence of a particular molecule, but rather by a distinctive **structural state**.

VIII. Comparing Bacteria with Animal and Plant Cells

Bacteria are generally much smaller than ordinary animal or plant cells. More importantly, bacteria themselves are a distinct type of cell — **prokaryotic cells**.

The most fundamental difference between bacteria (prokaryotes) and animal or plant cells (eukaryotes) lies in their degree of structural complexity.

Feature	**Bacteria (Prokaryotic Cells)**	**Animal / Plant Cells (Eukaryotic Cells)**
Nucleus	No true nucleus; DNA located in a nucleoid region	Membrane-bound nucleus containing DNA
Organelles	Very simple; ribosomes only	Highly complex; mitochondria, Golgi apparatus, endoplasmic reticulum, etc.
Cell Wall	Present (primarily peptidoglycan)	Absent in animals; present in plants (cellulose)
Biological Organization	Unicellular organisms	Predominantly multicellular organisms
DNA Structure	Usually circular	Usually linear

Table 2-10-4-1: Structural comparison between bacteria and animal/plant cells.

IX. Conclusion

Viruses and bacteria are more than two categories of microorganisms. They represent two successive stages in the emergence of life.

Viruses embody the earliest appearance of biological structural organization, whereas bacteria achieve the first complete structural loop of life.

Consequently, the transition from viruses to bacteria marks one of the most significant milestones in the progression from **chemical structural periodicity** to **biological structural periodicity**.

Within the Structural Periodic Table, this transition represents the **first complete integration of Point (•), Line (|), and Circle (O)** into an autonomous living system.

Closing Remark:

Viruses preserve fragments of life; bacteria complete the first living system. The true boundary of life is crossed not when molecules become more complex, but when matter, dynamics, and structural closure become a self-sustaining whole.

Section 5. Chapter Summary – From Chemistry to Life

This chapter has examined the four fundamental stages preceding and accompanying the emergence of life:

- **Point (•): Life Window** — the elemental foundation of life.
- **Line (|): Life Dynamic Chain** — self-sustaining reaction networks.
- **Circle (O): Life Structural Closure** — stable biological architectures, including DNA, cell membranes, and proteins.
- **Life:** The first complete living systems — represented by viruses and bacteria.

Together, these stages establish the following structural progression: **Elements → Reactions → Configurations → Life**

corresponding to **Point (•) → Line (|) → Circle (O) → Living Systems**

From this perspective, life did not emerge as a sudden event. Rather, it arose naturally as structural closure gradually developed through successive levels of organization.

Closing Remark:

Elements provide the materials of life; reaction networks provide its dynamics; molecular configurations provide its stability. When these three components were first unified in bacteria, chemistry crossed the boundary and became life.

As structural complexity increases, strict periodicity gradually becomes less prominent, while **structural recurrence** becomes increasingly significant.

The **Periodic Table of Elements** demonstrates the regularity of **periodicity**, whereas living systems demonstrate the regularity of **structure**.

In other words, as organization advances toward higher levels, periodic repetition becomes progressively weaker, while structural organization becomes increasingly dominant.

The **Structural Periodic Table** is proposed precisely to describe this hierarchical transition — from **periodic recurrence** toward **structural recurrence**.

This transition also forms the conceptual bridge between chemistry and biology. While chemistry primarily investigates the periodic organization of matter, life sciences increasingly reveal recurring structural principles operating across multiple hierarchical levels.

Accordingly, the next chapter moves beyond the chemical origin of life to investigate the **Structural Periodicity of Life**, exploring how living systems continually reproduce, amplify, and reorganize themselves through recurring structural principles.

Note: The complete reference list for this chapter appears below. The following three works are particularly relevant because they closely correspond to the central ideas developed in this chapter.

1) Forterre, P. (2010). ***Defining Life: The Virus Viewpoint.***

This work examines one of the fundamental questions addressed in Section 4: **Should viruses be regarded as living organisms?**

It provides an important perspective on the structural boundary between chemistry and life.

2) Woese, C. R. (1998). ***The Universal Ancestor.***

Woese investigates how the earliest living systems may have emerged. His discussion closely parallels the structural question raised in this chapter: **Why do bacteria represent the first complete living system?**

3) Maturana, H. R., & Varela, F. J. (1980). ***Autopoiesis and Cognition.***

This work is particularly significant because it proposes that the defining characteristic of life is not a particular substance, but the capacity of a system to maintain and reproduce itself continuously.

This conclusion closely parallels the structural interpretation developed in this chapter:

Life = Information + Dynamics + Boundary

Within the framework of the **Structural Periodic Table**, these correspond respectively to:

- **Point (•):** Life Window (material source)
- **Line (|):** Life Dynamic Chain
- **Circle (O):** Life Structural Closure

These three structural roles exhibit a strong conceptual correspondence with **autopoiesis**, understood as the continuous generation and self-maintenance of living systems.

Rather than replacing contemporary theories of life, the Structural Periodic Table seeks to provide a higher-level structural framework within which these biological principles may be interpreted in a unified manner.

Chemistry explains how matter becomes organized. Biology explains how organized matter becomes living. The Structural Periodic Table seeks to explain why this transition repeatedly occurs through the progressive closure of structure.

References (Chapter 10):

Biochemistry and Molecular Foundations of Life

[1] **Alberts, B., Johnson, A., Lewis, J., Morgan, D., Raff, M., Roberts, K., & Walter, P.** (2022). *Molecular Biology of the Cell* (7th ed.). Garland Science.

[2] **Berg, J. M., Tymoczko, J. L., Gatto, G. J., & Stryer, L.** (2019). *Biochemistry* (9th ed.). W. H. Freeman.

[3] **Lehninger, A. L., Nelson, D. L., & Cox, M. M.** (2021). *Lehninger Principles of Biochemistry* (8th ed.). W. H. Freeman.

[4] **Voet, D., Voet, J. G., & Pratt, C. W.** (2019). *Fundamentals of Biochemistry: Life at the Molecular Level* (6th ed.). John Wiley & Sons.

[5] **Lodish, H., Berk, A., Kaiser, C. A., et al.** (2021). *Molecular Cell Biology* (9th ed.). W. H. Freeman.

Origin of Life and Prebiotic Chemistry

[6] Schrödinger, E. (1944). *What Is Life?* Cambridge University Press.

[7] Miller, S. L. (1953). A Production of Amino Acids Under Possible Primitive Earth Conditions. *Science*, 117(3046), 528–529.

[8] Orgel, L. E. (1998). The Origin of Life — A Review of Facts and Speculations. *Trends in Biochemical Sciences*, 23(12), 491–495.

[9] de Duve, C. (1995). *Vital Dust: Life as a Cosmic Imperative.* Basic Books.

[10] Smith, E., & Morowitz, H. J. (2016). *The Origin and Nature of Life on Earth*. Cambridge University Press.

[11] Luisi, P. L. (2006). *The Emergence of Life: From Chemical Origins to Synthetic Biology*. Cambridge University Press.

[12] Maynard Smith, J., & Szathmáry, E. (1999). *The Origins of Life: From the Birth of Life to the Origin of Language.* Oxford University Press.

Autocatalysis, RNA World and Metabolic Networks

[13] Kauffman, S. A. (1986). Autocatalytic Sets of Proteins. *Journal of Theoretical Biology*, 119(1), 1–24.

[14] Kauffman, S. A. (1993). *The Origins of Order: Self-Organization and Selection in Evolution*. Oxford University Press.

[15] Dyson, F. J. (1985). *Origins of Life*. Cambridge University Press.

[16] Morowitz, H. J. (1968). *Energy Flow in Biology.* Academic Press.

[17] Morowitz, H. J. (1992). *Beginnings of Cellular Life: Metabolism Recapitulates Biogenesis*. Yale University Press.

[18] Gilbert, W. (1986). The RNA World. *Nature*, 319(6055), 618.

[19] Joyce, G. F. (2002). The Antiquity of RNA-Based Evolution. *Nature*, 418(6894), 214–221.

[20] Cech, T. R. (1986). A Model for the RNA-Catalyzed Replication of RNA. *Proceedings of the National Academy of Sciences*, 83(12), 4360–4363.

[21] Gesteland, R. F., Cech, T. R., & Atkins, J. F. (2006). *The RNA World* (3rd ed.). Cold Spring Harbor Laboratory Press.

Biological Structure and Self-Organization

[22] Watson, J. D., & Crick, F. H. C. (1953). Molecular Structure of Nucleic Acids: A Structure for Deoxyribose Nucleic Acid. *Nature*, 171(4356), 737–738.

[23] Singer, S. J., & Nicolson, G. L. (1972). The Fluid Mosaic Model of the Structure of Cell Membranes. *Science*, 175(4023), 720–731.

[24] Anfinsen, C. B. (1973). Principles That Govern the Folding of Protein Chains. *Science*, 181(4096), 223–230.

[25] Dobson, C. M. (2003). Protein Folding and Misfolding. *Nature*, 426(6968), 884–890.

[26] Maturana, H. R., & Varela, F. J. (1980). *Autopoiesis and Cognition: The Realization of the Living*. D. Reidel Publishing Company.

[27] Deacon, T. W. (2012). *Incomplete Nature: How Mind Emerged from Matter*. W. W. Norton & Company.

[28] Prigogine, I., & Stengers, I. (1984). *Order Out of Chaos*. Bantam Books.

[29] Holland, J. H. (1998). *Emergence: From Chaos to Order*. Oxford University Press.

[30] Mitchell, M. (2009). *Complexity: A Guided Tour*. Oxford University Press.

[31] Barabási, A.-L. (2016). *Network Science*. Cambridge University Press.

Viruses, Bacteria and the Boundary of Life

[31] Lwoff, A. (1957). The Concept of Virus. *Journal of General Microbiology*, 17(2), 239–253.

[32] Villarreal, L. P. (2004). *Viruses and the Evolution of Life*. ASM Press.

[33] Koonin, E. V., Senkevich, T. G., & Dolja, V. V. (2006). The Ancient Virus World and Evolution of Cells. *Biology Direct*, 1(1), 29.

[34] Claverie, J.-M. (2006). Viruses Take Center Stage in Cellular Evolution. *Genome Biology*, 7(6), 110.

[35] Forterre, P. (2010). Defining Life: The Virus Viewpoint. *Origins of Life and Evolution of Biospheres*, 40(2), 151–160.

[36] Madigan, M. T., Bender, K. S., Buckley, D. H., Sattley, W. M., & Stahl, D. A. (2021). *Brock Biology of Microorganisms* (16th ed.). Pearson.

[37] Woese, C. R. (1998). The Universal Ancestor. *Proceedings of the National Academy of Sciences*, 95(12), 6854–6859.

[38] Cleland, C. E., & Chyba, C. F. (2002). Defining Life. *Origins of Life and Evolution of the Biosphere*, 32(4), 387–393.

Structural Periodicity and Metastructural Framework

[39] Chang, J. (2013). *Great Ultimate Theory*. Universal Publishing.

[40] Chang, J. (2026). *Metastructural Unification (Chemistry Volume): Unifying Reactions, Compounds and Elements*. Universal Publishing.

Chapter Eleven: Life Structural Periodicity

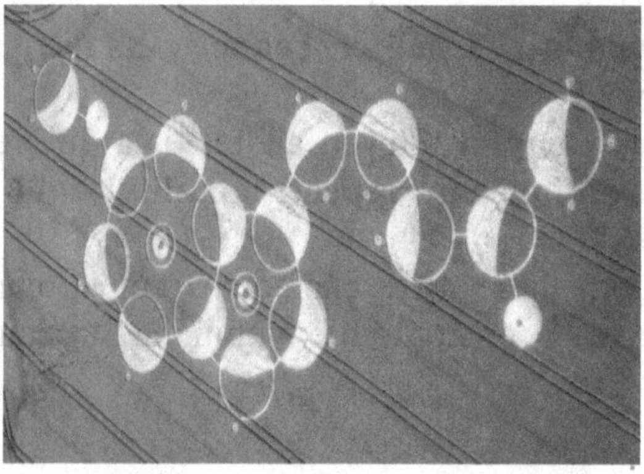

Figure 2-11: A crop circle illustrating Galactic Civilization. For details, see《Crop Circle》.

Chapter Overview:

The triadic structure of life corresponds to the three fundamental structural roles of **Line (|), Circle (O), and Point (•)**, forming the common foundation of structural periodicity throughout the natural sciences.

This chapter examines life from the perspective of the **Structural Periodic Table**. It proposes that life is not an independent system existing outside the natural sciences, but rather the long-term recursive outcome of mathematics, physics, and chemistry. Based on this perspective, five fundamental laws of life are proposed to explain why living systems are able to emerge, continually expand, and sustain themselves through successive structural organization.

Chapter Ten (End of the Chemistry Section): Life Window → Life Dynamic Chain → Life Structural Closure → First Complete Living System (Virus → Bacterium)

Chapter Eleven (Beginning of the Life Section): Cell → Structural Periodicity → Structural Recursion → Triadic

Functional Complementarity → Hierarchical Transition → Structural Mapping

In other words, **Chapter Ten** answers the question: **How does life emerge?**

whereas **Chapter Eleven** addresses a different question: **Why does life continue to repeat, amplify, and evolve after its emergence?**

Together, these two chapters form a complete theoretical transition from chemistry to life within the Structural Periodic Table.

Section 1. Structural Periodicity Law – The Cell as the First Stable Structural Closure of Life

Abstract:

Previous chapters have discussed structural periodicity in mathematics, physics, and chemistry, and further examined the structural characteristics of biological precursors such as viruses, DNA, and RNA. Although these systems already possess certain capacities for information organization, they are still unable to complete an independent life cycle.

Modern biology regards the **cell** as the fundamental structural and functional unit of life. From the perspective of the **Structural Periodic Table**, however, the significance of the cell extends beyond being merely the smallest living unit. More importantly, it represents the first successful integration of the three fundamental structural functions: **Generation (Point •) → Evolution (Line ｜) → Closure (Circle O)**

For the first time, life became a stable, self-maintaining system capable of self-replication and continuous evolution.

Therefore, the cell is not only the starting point of modern biology but also the **first stable structural closure** after structural periodicity enters the biological level, leading to what we propose as the **Law of Structural Periodicity**.

I. Why Does the Cell Represent the First Stable Structural Closure of Life?

A typical eukaryotic cell consists of three principal components:

- **Nucleus**
- **Cytoplasm**
- **Cell Membrane**

Traditional biology usually explains these structures independently according to their biological functions.

From the viewpoint of the **Structural Periodic Table**, however, they correspond precisely to the three fundamental structural dynamics.

1.1 Point (•): The Nucleus – The Center of Biological Generation

The nucleus stores the genetic information encoded in DNA. It is responsible for:

- preservation of genetic information;
- DNA replication;
- regulation of gene expression.

Accordingly, the nucleus corresponds to the **Point (•)** — the local source of biological generation.
Without the nucleus, life loses its informational center.

1.2 Line (|): The Cytoplasm – The System of Biological Evolution

The cytoplasm contains numerous organelles, including:

- ribosomes;
- endoplasmic reticulum;
- Golgi apparatus;
- mitochondria;
- lysosomes.

Together they continuously perform:

- energy conversion;
- intracellular transport;
- protein synthesis;
- information transmission;
- metabolism.

Therefore, the cytoplasm corresponds to the **Line (|)**, representing the continuous evolutionary processes occurring within living systems.

Life is never static—it exists through constant structural activity.

1.3 Circle (O): The Cell Membrane – Stable Structural Closure

The greatest significance of the cell membrane is not simply that it surrounds the cell.

Rather, it establishes the distinction between the internal and external environments. The membrane is responsible for:

- regulating material exchange;
- maintaining internal stability;
- establishing the living environment;
- preserving systemic integrity.

Accordingly, it corresponds to the **Circle (O)** — the first genuine stable structural closure of life. Without the cell membrane, no independent living system could exist.

II. The Triadic Structure Forms a Living System for the First Time

For the first time, the cell simultaneously possesses all three structural roles.

Structural Role	Biological Component	Primary Function
Point (•)	Nucleus	Information generation
Line (|)	Cytoplasm	Biological evolution
Circle (O)	Cell membrane	Stable structural closure

Table 2-11-1-1: The Structural Closure Cycle of the Cell

Together they establish the structural sequence: **Point → Line → Circle**

Or **Information Generation → Functional Evolution → Systemic Closure**

This structural cycle enables life for the first time to exhibit:

- **Self-organization**
- **Self-maintenance**
- **Self-replication**
- **Self-evolution**

Accordingly, the cell becomes the genuine starting point of biological structure.

III. Why Can Cells Persist Over Long Periods?

Viruses are capable of replication only through host cells.

Cells, however, are able to:

- obtain energy;
- regulate their internal environment;
- repair damage;
- divide;
- maintain homeostasis.

Together these capabilities establish the complete structural periodic cycle of life.

The cell is therefore the first system to realize: **Generation → Evolution → Closure → Regeneration**

that is, **new cells.** Life thus enters true recursion.

From the perspective of structural periodicity, this marks the first stable recursive cycle in biological organization.

IV. The Position of the Cell within the Structural Periodic Table

Arranged hierarchically, biological organization may be represented as: **Quarks → Protons & Neutrons → Atoms → Molecules → Nucleotides → DNA/RNA → Viruses → Cells → Multicellular Organisms → Complex Living Systems**

All preceding structural levels ultimately converge upon the cell. Modern biology identifies the cell as the basic unit of life.

The **Structural Periodic Table** further proposes that the cell is also the **first living system to complete triadic structural closure**.

Beyond storing hereditary information, it establishes:

- a mechanism of generation;
- a mechanism of evolution;
- a mechanism of structural closure.

Accordingly, the true emergence of life is marked neither merely by DNA nor by replication itself, but by the formation of a stable self-maintaining structural system.

The cell therefore represents both the culmination of chemical structure and the beginning of biological structure. It serves as the fundamental bridge connecting the natural sciences with the life sciences.

V. Conclusion

Modern biology recognizes the cell as the fundamental unit of life and the classical example of a complete biological life cycle.

From the perspective of the **Structural Periodic Table**, however, the cell possesses an even deeper significance: it is the first system to achieve the complete integration of the three fundamental structural dynamics:

- **Point (•)** — generation of biological information;
- **Line (|)** — evolution of biological functions;
- **Circle (O)** — stable structural closure of living systems.

The emergence of the cell therefore marks not only the beginning of life itself, but also the first successful transition of structural periodicity from the material world to the living world.

Closing Statement: Viruses initiated biological information, whereas cells completed biological closure. From that moment onward, structural periodicity ceased to be merely the repetition of material structures and became the foundation of life's capacity for self-generation, self-maintenance, and self-evolution.

Life does not begin with genetic information alone; it begins when generation, evolution, and structural closure become integrated into a complete self-maintaining system.

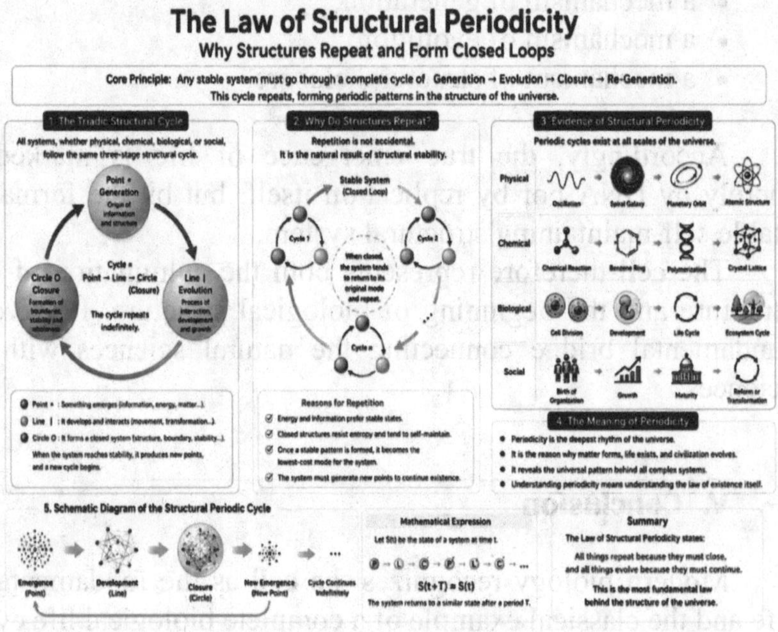

Figure 2-11-1-1: The Law of Structural Periodicity: Why Do Living Structures Repeat?

Section 2. Structural Recursion Law – The First Demonstration of Structural Recursion from Quarks to Cells

Abstract:

The previous chapter discussed the structural evolution from elements, molecules, viruses, and finally to cells. This section proposes a further insight: life is not merely an accumulation of isolated structures, but the continuous recursion of the same structural pattern across different hierarchical levels.

From quarks, protons, atoms, nucleotides, DNA, viruses, and ultimately cells, one can observe a recurring organizational principle. Although each level consists of fundamentally different physical entities, they repeatedly exhibit similar triadic organizations while giving rise to increasingly higher structural hierarchies.

This recurring phenomenon is referred to here as **Structural Recursion**.

If this principle is valid, then the origin of life should not be regarded as a single accidental event, but rather as the natural consequence of the continual recursive unfolding of universal structure.

I. Life Did Not Appear Suddenly

Modern life science primarily studies: **DNA, RNA, Genes, Viruses, Cells.**

However, tracing these structures further downward reveals that life did not emerge suddenly. Instead, it developed from increasingly fundamental material structures.

Its developmental pathway may therefore be represented as:
Quarks → Protons and Neutrons → Atoms → Nucleotides → DNA/RNA → Viruses → Cells

These are not seven unrelated entities. Rather, they constitute one continuous chain of structural generation.

II. Quarks: The First Triadic Structure

Modern particle physics shows that a proton consists of:

- two **up quarks**, and
- one **down quark**.

A neutron consists of:

- two **down quarks**, and
- one **up quark**.

Every nucleon is therefore itself a triadic combination.

This relationship may be expressed as: **Quarks → Three-Body Structure → Proton / Neutron**

From the perspective of the **Structural Periodic Table**, quarks represent the first appearance of stable existence through triadic organization. This is also the earliest manifestation of structural recursion in nature.

III. Atoms: The Second Recursion of Triadic Structure

Protons and neutrons combine to form atomic nuclei, while electrons orbit around the nucleus, giving rise to atoms.

This structure may be abstracted as: **Atomic Nucleus → Electron Motion → Stable Electron Shells**

Corresponding to: **Point • → Line | → Circle O**

For the first time, the complete sequence of **Generation → Propagation → Stabilization**

appears. Atoms therefore represent the second recursive realization of the triadic structure.

IV. Nucleotides: The First Emergence of Biological Information

Nucleotides are not random molecular assemblies.

Each nucleotide consists of three principal components:

- a pentose sugar,
- a phosphate group,
- a nitrogenous base.

Together they form one complete nucleotide. In structural language: **Sugar → Phosphate Linkage → Base Encoding,**

forms a new triadic system. More importantly, nitrogenous bases begin to carry informa-tional functions. For the first time, life possesses information that can be:

- encoded,
- replicated,
- inherited.

Consequently, nucleotides accomplish not only molecular assembly but also the generation of biological information.

V. DNA: Triadic Structure Becomes Recursive Information

DNA is far more than a large molecule. It performs three essential functions:

1) Storage of genetic information;
2) Guidance of protein synthesis;
3) Replication and inheritance.

This may be summarized as: **Information → Expression → Replication**

Corresponding to: **Point • → Line | → Circle O**

DNA is the first biological structure capable of continuous structural recursion.

Biological information is now able to persist across time.

VI. Viruses: Structural Recursion at the Boundary of Life

Viruses consist primarily of:

- nucleic acids,
- protein capsids.

Although incapable of independent life, viruses are nevertheless able to:

- replicate,
- mutate,
- evolve.

Viruses therefore represent structural recursion approaching the boundary of life. Although they have not yet become complete living systems, they already possess the cycle of: **Information → Replication → Evolution**

Consequently, viruses form a critical transitional level between chemical structures and biological structures.

VII. Cells: The First Complete Structural Recursion

Only with the emergence of cells does structural recursion become fully closed. The cell itself may be represented as: **Nucleus → Cytoplasm → Cell Membrane**

These correspond respectively to:

- genetic information,
- metabolism,
- structural boundary.

Thus, **Point • → Line | → Circle O**

forms: **Generation → Evolution → Structural Closure → New Cell**

For the first time, this entire cycle becomes self-sustaining.

Cells therefore constitute the first fully recursive structural system of life.

VIII. Unified Observation: One Structure Repeated Continuously

When the entire developmental sequence is viewed together, the following correspondence emerges:

Hierarchical Level	Point •	Line |	Circle O
Quark	Up / Down Quarks	Three-quark Combination	Proton / Neutron
Atom	Atomic Nucleus	Electron Motion	Stable Electron Shell
Nucleotide	Pentose Sugar	Phosphate Linkage	Base Encoding
DNA	Genetic Information	Gene Expression	Stable Double Helix
Virus	Nucleic Acid	Infection and Replication	Protein Capsid
Cell	Nucleus	Cytoplasmic Metabolism	Cell Membrane

Table 2-11-2-1. Structural Recursion from Quarks to Cells

The physical objects continuously change. The structure continually repeats. Life therefore does not continuously invent entirely new structures. Rather, it continuously **recursively reorganizes existing structural principles.**

IX. Reinterpretation within the Structural Periodic Table

Traditional life science explains life through long-term biological evolution.

The **Structural Periodic Table** further proposes that biological evolution continues because the triadic structure continually undergoes recursion.

Accordingly, what persists throughout evolution is not any particular molecule, but the structural cycle of: **Generation → Evolution → Structural Closure**

Every higher level of biological organization is simply another recursive realization of this same structural cycle.

X. Conclusion

From quarks to cells, one observes the repeated emergence of the same structural organization:

Quarks form nucleons. Nucleons form atoms. Atoms form nucleotides. Nucleotides form DNA. DNA gives rise to viruses. Viruses ultimately develop into cells.

This continuous sequence suggests that life did not appear abruptly. Rather, it emerged naturally through successive stages of structural recursion.

Therefore, what life truly inherits is not a particular substance, but a structural pattern capable of continual generation, evolution, and stable closure.

Structural Conclusion:

From quarks to cells, matter changes, scale changes, and function changes, yet the underlying pattern of structural recursion remains remarkably consistent.

The continuity of life may therefore originate not from any particular molecule, but from the continual repetition of the same generative structural pattern.

Any stable structure emerging at a higher hierarchical level is fundamentally the recursive reorganization of the triadic structure—Point •, Line ｜, and Circle O—within a new level of organization.

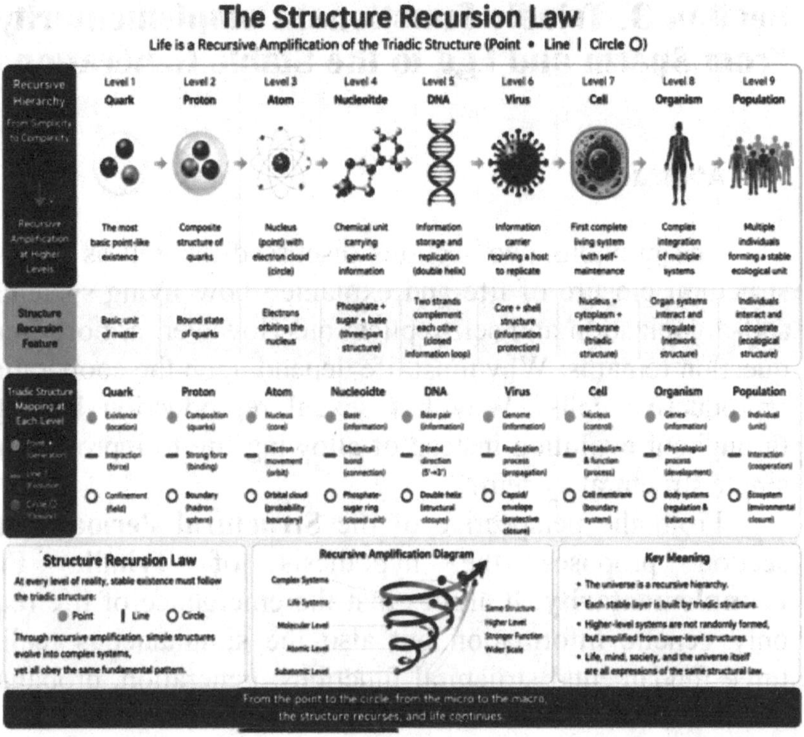

Figure 2-11-2-1: Structural Recursion Law — Why Structures Continuously Evolve Toward Higher Levels

Section 3. Triadic Functional Complementarity Law – From Sperm and Egg to the Stable Generation of Life

Abstract:

The previous two sections discussed the cell as the first stable structural closure of life and explained how living systems achieve self-organization and self-replication. However, a more fundamental question remains: Why must life depend upon the cooperation of two reproductive cells? Why has sexual reproduction been preserved throughout evolution instead of allowing life to reproduce through a single structural system?

From the perspective of the **Structural Periodic Table**, this section proposes the hypothesis of **Triadic Functional Complementarity**. It argues that the emergence of life requires not only genetic information but also the simultaneous realization of three fundamental structural functions: generation, propagation, and stabilization.

Sperm and egg perform different structural roles. Only when fertilization completes the closure of these three structural functions can a new living system truly begin. Consequently, the origin of life should be understood not merely as the union of genetic material, but as the first complete realization of complementary triadic structural functions.

I. Why Can Life Not Arise from a Single Structure?

Nature contains many forms of asexual reproduction, including:

- bacterial fission,
- yeast budding,
- vegetative propagation in plants.

However, the more complex an organism becomes, the more it depends upon **sexual reproduction**. Why?

Traditional biology explains this primarily through increased genetic diversity. The **Structural Periodic Table** offers a further interpretation: Life requires not only genetic variation, but also **functional structural complementarity**.

The essential question is therefore not: **Why are two reproductive cells required?**

Rather: **Why must life achieve the closure of a complete triadic structure?**

II. The Structural Role of Sperm: A Line-Dominated Generative System

The defining characteristic of sperm is not its small size, but its extraordinary mobility. Its principal functions include:

- actively locating the egg,
- transporting genetic information over long distances,
- initiating fertilization.

Its dominant structural role therefore corresponds to the **Line** (|), representing:

- propagation,
- connection,
- dynamic evolution,
- initiation.

Naturally, sperm also carries complete DNA and therefore retains the characteristics of the **Point**.

Overall, however, its structural identity is best described as: **Point + Line**

representing a **local generative source combined with dynamic propagation.**

III. The Structural Role of the Egg: A Circle-Dominated Stabilizing System

The egg performs an almost opposite function. It:

- exhibits minimal movement,
- stores abundant nutrients,
- supplies essential organelles,
- provides the developmental environment.

Following fertilization, the early embryo depends almost entirely upon the egg. Its dominant structural role therefore corresponds to the **Circle (O)**, representing:

- stability,
- containment,
- structural closure,
- developmental support.

Like sperm, the egg also contains DNA and therefore retains the characteristics of the **Point**.

Its overall structural identity is therefore: **Point + Circle**

representing a **local generative source combined with a stable structural environment.**

IV. Fertilization: The First Complete Closure of the Triadic Structure

When the sperm enters the egg, life achieves, for the first time, the complete integration of: **Point • — Line | — Circle O**

At this stage:

• **Point •** originates from the combined genetic information of both parents.

• **Line |** is represented by sperm motility, transport, fusion, and activation.

• **Circle O** is provided by the egg through its stable environment, membrane, and nutrient reserves.

Thus the complete structural sequence emerges: **Point → Line → Circle**

forming the first complete structural cycle of life.

Fertilization therefore signifies not merely the combination of genetic material, but the first successful closure of the complete triadic structural system.

V. Why Can Only the Complete Triadic Structure Produce Life?

If only **Point + Circle** exist, without propagation, life cannot begin.

If only **Point + Line** exist, without stability, life cannot persist.

Without the **Point,** no genetic information exists.

Accordingly, the absence of any one structural component interrupts the entire life cycle. This leads to the proposed **Law of Triadic Functional Complementarity**.

A living system must simultaneously possess:

- **Information Generation (Point)**
- **Dynamic Propagation (Line)**
- **Structural Stability (Circle)**

Only through the integration of all three functions can life achieve long-term existence.

VI. Triadic Functional Complementarity Extends Beyond Fertilization

This principle is not confined to reproduction.

It repeatedly appears throughout biological organization.

Hierarchical Level	**Point •**	**Line ǀ**	**Circle O**
DNA	Genetic Code	Self-Replication	Stable Double Helix
Cell	Nucleus	Cytoplasm	Cell Membrane

Reproduction	Sperm DNA	Sperm Motility	Egg Environment
Organism	Genome	Growth and Development	Physiological Homeostasis

Table 2-11-3-1: Triadic Functional Complementarity Law

Triadic functional complementarity is therefore not an isolated biological phenomenon.

It may instead represent one of the fundamental structural patterns repeatedly organizing living systems.

VII. Conclusion

Modern biology generally regards fertilization as the union of two genetic materials.

From the perspective of the **Structural Periodic Table**, however, its deeper significance lies in achieving, for the first time, the complete integration of the three fundamental structural functions:

- **Point** • provides genetic information.
- **Line** | provides propagation and activation.
- **Circle O** provides stability and structural closure.

Together these establish the structural foundation for continual generation, self-maintenance, and long-term biological evolution.

Consequently, the true continuity of life depends not merely upon genetic replication, but upon the stable closure created through **Triadic Functional Complementarity**.

Structural Conclusion: Life does not arise from a single structure. It emerges through the complementarity of three fundamental structural functions.

What truly gives birth to life is not DNA alone, but the first successful integration of **generation, propagation, and stabilization** into one complete structural system.

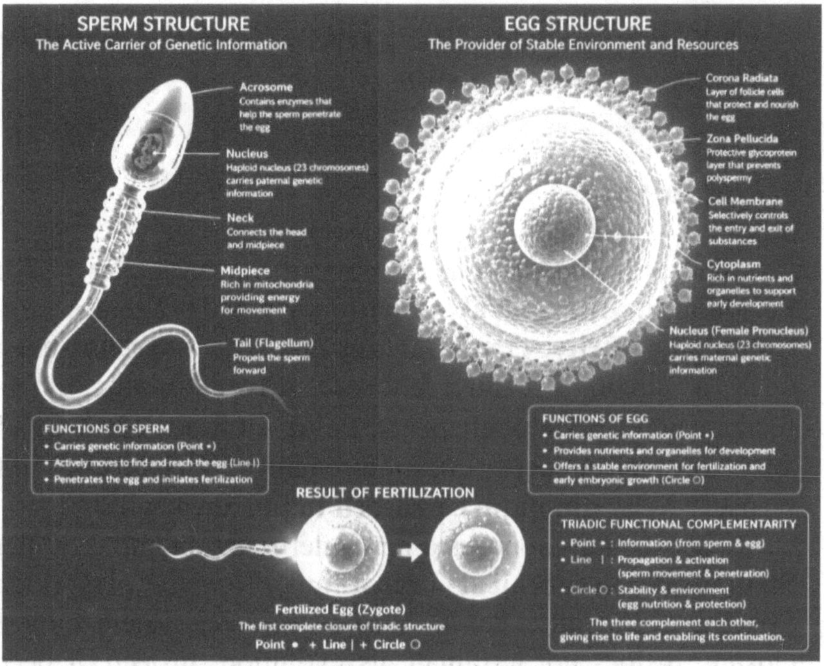

Figure 2-11-3-1: Triadic Functional Complementarity Law — Why Life Becomes Possible

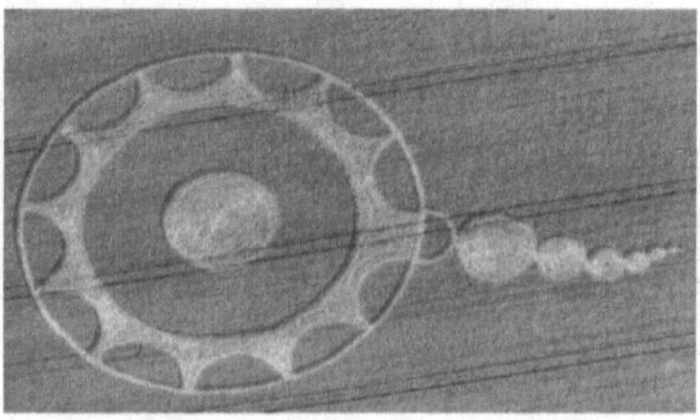

Figure 2-11-3-2: Fertilization of Sperm and Egg (see *Crop Circle Civilization*).

This principle also provides a natural theoretical bridge to the subsequent volume, **Hexagonal Structure**, where the same triadic functional organization continues to appear in life systems, animals, plants, psychology, and other higher-level structures, forming a unified structural framework.

Section 4. Structural Hierarchical Transition Law – From Viruses to Cells

Abstract:

The previous sections introduced the principles of **Structural Periodicity**, **Structural Recursion**, and **Triadic Functional Complementarity**, demonstrating that living systems are not built upon a single structure but are driven jointly by the three fundamental structural elements: **Point •, Line ｜, and Circle O**.

However, a deeper question remains: **Why did life not stop at viruses? Why did nature proceed to produce cells?**

This section proposes a new interpretation based on the **Law of Structural Hierarchical Transition**. From the perspective of the **Structural Periodic Table**, once a structural system achieves stable closure, it does not terminate its evolution. Instead, it becomes the starting point for a new and higher structural level.

Viruses therefore do not represent the endpoint of biological evolution; rather, they constitute a stable structural window preceding the emergence of cells.

Life advances from simple to increasingly complex organization through the continual repetition of the structural cycle: **Generation → Evolution → Structural Closure**

thus producing successive hierarchical transitions.

I. What Is Structural Hierarchical Transition?

Traditional biology primarily investigates:

- how genes replicate,
- how DNA is expressed,
- how cells divide.

The **Structural Periodic Table** raises a different question: **Why does life continually progress toward higher structural levels?**

If life merely reproduced itself indefinitely, why did increasingly complex forms emerge, such as:

- viruses,
- prokaryotic cells,
- eukaryotic cells,
- multicellular organisms?

Life therefore develops not simply through replication, but through continual **structural hierarchical transition**.

II. Viruses: The First Stable Structural Window of Life

Viruses possess the simplest biological structures.
Their principal components include:

- nucleic acids (DNA or RNA),
- protein capsids.

In structural language:

- **Point •** — Genetic information (DNA/RNA)
- **Line ｜** — Replication process
- **Circle O** — Protective protein capsid

Viruses therefore already exhibit the minimal structural cycle:
Point → Line → Circle
However, viruses still lack:

- independent metabolism,
- autonomous energy production,
- self-sustaining reproduction.

Consequently, they remain dependent upon host cells for survival. Viruses therefore accomplish the **first stable structural closure**, but they have not yet achieved the **first structural hierarchical transition**.

III. Cells: The First Genuine Hierarchical Transition

Compared with viruses, cells introduce entirely new structural systems, including:

- the cell membrane,
- the cytoplasm,
- the nucleus.

Together these establish a new triadic organization:

- **Point •** — Nucleus (genetic information)
- **Line** | — Cytoplasm (metabolic network)
- **Circle O** — Cell membrane (structural boundary)

Whereas viruses possess only: **Genetic Information + Protective Structure**

cells possess: **Genetic Information → Metabolism → Structural Boundary**

forming the first independently functioning living system.

The transition from **virus to cell** therefore represents not merely an increase in complexity or size, but a genuine **structural hierarchical transition**.

IV. Why Does Hierarchical Transition Occur?

According to the **Structural Periodic Table**, once one structural level reaches stable closure, new functional possibilities emerge naturally.

For example: **Nucleic Acids → Viruses → Cells → Multicellular Organisms → Tissues → Organs → Complete Organisms**

Each transition represents not simply quantitative growth, but the emergence of an entirely new stable organizational level.

Hierarchical transition is therefore driven by **the formation of new structural closure**.

V. Structural Hierarchical Transition Is Recursive

Biological evolution repeatedly follows the same structural sequence: **Local Generation → Structural Propagation → Global Structural Closure → A New Hierarchical Level**

This progression may be illustrated as: **Quark → Proton → Atom → Molecule → Nucleotide → Gene → DNA/RNA → Virus → Cell → Multicellular Life → Organism**

Although each successive level becomes increasingly complex, the underlying structural mechanism remains fundamentally unchanged.

Life therefore does not continually invent new structural rules.

Instead, it continually repeats the same structural language at progressively higher levels.

VI. Reinterpretation within the Structural Periodic Table

Within the framework of the **Structural Periodic Table**, biological evolution may be represented as:

Point • → Generates new structural units → **Line** | → Propagates and develops these structures → **Circle O** → Forms a new stable structural system → A **new Point** • → Initiates the next hierarchical level.

Accordingly, biological evolution results from the combined action of:

- **Structural Recursion**, and
- **Structural Hierarchical Transition**.

VII. Conclusion

The development of life is not a continuous and gradual process alone, but proceeds through a series of stable structural windows.

Viruses represent the first stable structural closure of life.

Cells represent the first genuine structural hierarchical transition.

As each structural cycle repeatedly completes: **Generation → Evolution → Structural Closure**
new organizational levels continue to emerge.

Consequently, the increasing complexity of life results not merely from the accumulation of matter, but from the continual **upgrading of structural organization**.

Structural Conclusion: The evolution of life is not simply the accumulation of biological material. Rather, it is the continual realization of **structural hierarchical transitions.**

Every stable structural closure becomes the foundation upon which the next level of biological organization is constructed.

Structural Recursion explains why structures repeat; Structural Hierarchical Transition explains why structures ascend.

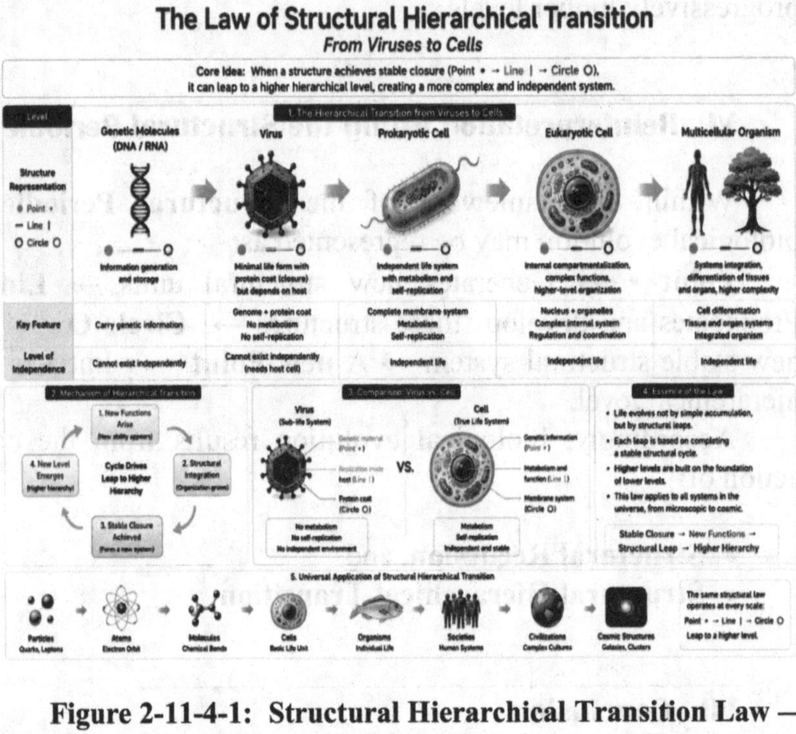

Figure 2-11-4-1: Structural Hierarchical Transition Law — Why Structures Progress to Higher Levels (e.g., Chemistry → Life → Civilization)

Section 5. Structural Recursive Mapping Law – From Genetic Information to the Amplification of Life

Abstract:

The previous sections introduced the principles of **Structural Periodicity**, **Structural Recursion**, **Triadic Functional Complementarity**, and **Structural Hierarchical Transition**, demonstrating that the development of life is not simply the accumulation of matter, but the continual completion of new structural closures.

This section further proposes a new structural perspective: **the development of life is fundamentally a process of continuous Structural Mapping.**

Beginning with DNA, genetic information is recursively amplified through cells, tissues, organs, and organ systems until a complete organism is formed. Subsequently, through the male and female reproductive systems, this complex biological organization is compressed once again into new genetic information, initiating another life cycle.

Life is therefore both a process of continual structural amplification and continual structural return to its informational origin.

From the perspective of the **Structural Periodic Table**, this recurring cycle — **Information → Structure → Life → Information**, constitutes the fundamental mechanism of **Structural Recursive Mapping**.

I. Why Can Life Continue to Expand?

Traditional biology primarily explains:

- how DNA replicates,
- how RNA is transcribed,
- how proteins are synthesized.

The **Structural Periodic Table** asks a different question: **How can a DNA molecule only a few nanometers in length eventually produce an entire human being?**

DNA occupies only nanoscopic dimensions, whereas the human body is billions of times larger.

The essential problem is therefore not molecular synthesis alone, but **cross-hierarchical structural amplification.**

II. The First Stage of Biological Amplification: Genetic Information

Life begins with DNA. DNA stores the genetic information that determines:

- body structure,
- protein composition,
- developmental programs,
- reproductive capability.

DNA therefore corresponds to the **Point (•)**.

It serves as the informational source of the entire living system. Although DNA does not directly construct the body, it determines how the body is organized.

Thus, the **Point** represents the origin of biological information.

III. The Second Stage of Biological Amplification: Cellular Execution

DNA alone cannot generate life.

Its information must pass through the sequence: **DNA → RNA → Protein → Cellular Metabolism → Cellular Function**

Cells therefore:

- receive genetic information,
- synthesize proteins,
- establish metabolic networks,
- perform biological activities.

Cells correspond primarily to the **Line (|)**. They execute:

- information transmission,
- functional implementation,
- biological operation.

IV. The Third Stage of Biological Amplification: Tissues and Organs

Large numbers of cells organize into tissues, including:

- nervous tissue,
- muscular tissue,
- epithelial tissue,
- connective tissue.

These tissues subsequently develop into organs such as:

- the heart,
- lungs,
- liver,
- kidneys,
- brain.

The coordinated interaction among these organs produces an integrated organism.

Organ systems therefore correspond to the **Circle (O)** because they establish:

- structural closure,
- systemic coordination,
- long-term stability.

Thus the sequence: **Point → Line → Circle**
achieves the first complete biological structural cycle.

V. The Reproductive System: Structural Mapping Begins Again

Life does not terminate with a single organism. Nature establishes both:

- the male reproductive system,
- the female reproductive system.

Together they initiate another cycle of structural mapping.

5.1 The Male Reproductive System

The male system is primarily responsible for:

- sperm production,
- genetic transmission,
- reproductive output.

Its dominant structural characteristic corresponds to the **Line (|)**.

The sequence: **Testis → Epididymis → Vas Deferens → Penis** primarily performs:

- transmission,
- propagation,
- connection.

Accordingly, the male reproductive system is best characterized as: **Point + Line**

5.2 The Female Reproductive System

The female system is primarily responsible for:

- ovum production,
- embryonic development,
- uterine protection,
- formation of new life.

Its dominant structural characteristic corresponds to the **Circle (O)**.

The sequence: **Ovary → Fallopian Tube → Uterus → Placenta**, creates a complete developmental environment.

Accordingly, the female reproductive system is best characterized as: **Point + Circle**

5.3 Closure of the Triadic Structure

A new living system emerges only through the union of: **Sperm (Point + Line)** and **Egg (Point + Circle) → Fertilization → New DNA → New Life**

Thus, **Point → Line → Circle**, is completed once again.

Life therefore proceeds not simply through DNA replication, but through the repeated closure of **Triadic Functional Complementarity**.

VI. Biological Mapping Exhibits Recursive Amplification

The complete developmental sequence may be represented as: **DNA → Gene → RNA → Protein → Cell → Tissue → Organ → Organ System → Organism → Male + Female → New DNA → Next Generation**

Every level simultaneously serves as:

- the amplification of the previous level,
- the foundation of the next level.

Life therefore develops not through linear growth, but through **Structural Recursive Mapping**.

VII. Reinterpretation within the Structural Periodic Table

From the perspective of the **Structural Periodic Table**, life continually proceeds through: **Information Generation → Functional Propagation → Structural Closure → Renewed Generation** forming the complete cycle: **DNA → Cell → Organ → Individual → Reproduction → DNA**

Accordingly, what continually expands throughout evolution is not matter itself, but **structure**.

VIII. Conclusion

The development of life is not a random accumulation of biological components. Rather, it is a continuous process of **Structural Mapping** and **Recursive Amplification**.

DNA provides genetic information. Cells execute biological functions. Tissues and organs establish integrated structural closure. The male and female reproductive systems initiate the next generation.

Life may therefore be summarized as the recurring sequence: **Genetic Information → Functional Execution → Complete Organism → Genetic Information**

This continuous cycle constitutes one of the fundamental mechanisms underlying biological evolution.

Hierarchical Level	Structural Role	Point •	Line ∣	Circle O
DNA	Genetic Information	✓		
RNA / Protein	Information Execution		✓	
Cell	Functional Unit	✓	✓	✓
Tissue	Functional Integration		✓	✓
Organ	Stable Biological Structure			✓

Organ System	Coordinated Organization	✓	✓	✓
Organism	Complete Life System	✓	✓	✓
Male Reproductive System	Propagation	✓	✓	
Female Reproductive System	Nurturing & Closure	✓		✓
Fertilization	New Structural Beginning	✓	✓	✓

Table 2-11-5-1. Structural Recursive Mapping in Biological Systems

Structural Conclusion: Life is not merely the amplification of matter. It is the recursive mapping of structure.

From DNA to the complete organism, and from the organism back to DNA, the same triadic structural cycle repeatedly appears across every hierarchical level of life.

Note:

This section no longer asks merely **how life is formed**. Instead, it proposes a unified structural framework explaining how genetic information is continually amplified through structural mapping and subsequently compressed back into genetic information through the reproductive system.

Together with the previous four sections —

- Structural Periodicity,
- Structural Recursion,
- Triadic Functional Complementarity,
- Structural Hierarchical Transition—

it establishes a comprehensive theoretical framework for biological structural organization.

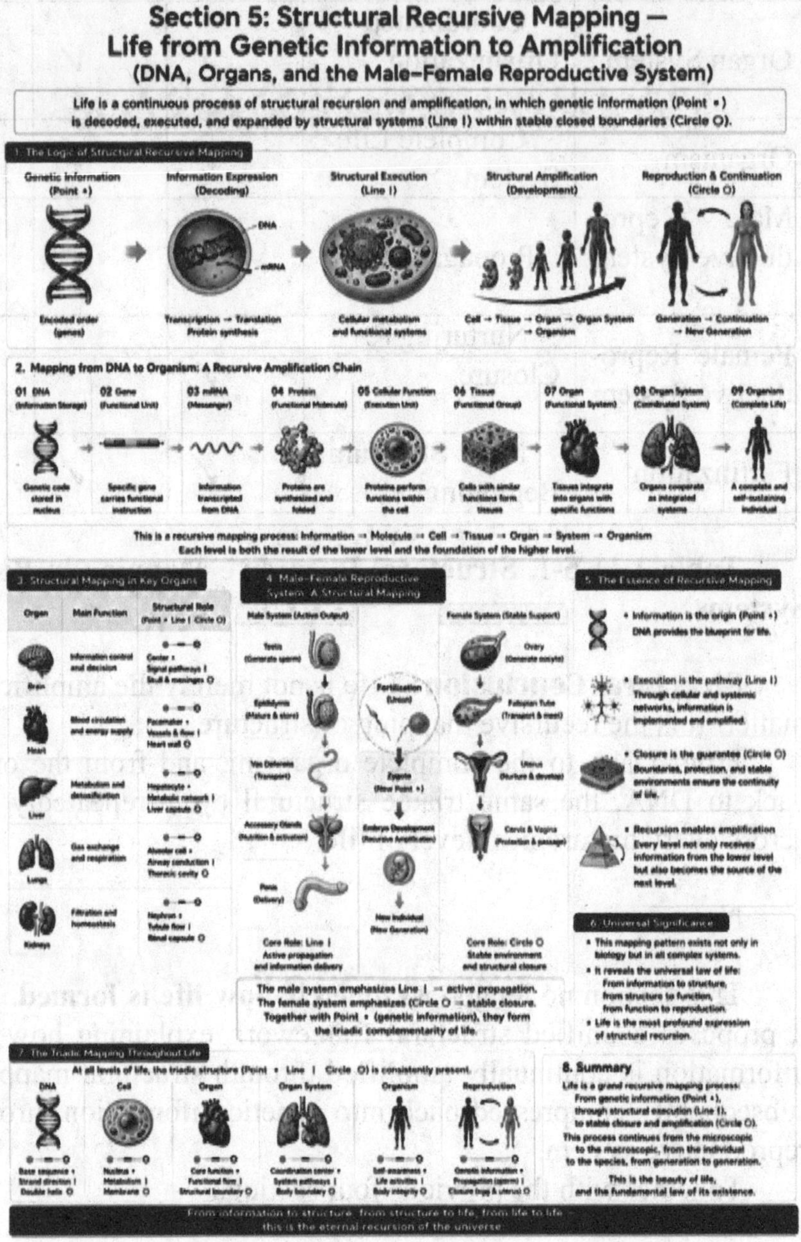

Figure 2-11-5-2: Structural Recursive Mapping Law — Why Higher Levels Increasingly Resemble Lower Levels

Furthermore, the correspondence between **sperm ↔ male reproductive system** and **egg ↔ female reproductive system** represents only the first piece of evidence.

If the proposed law is valid, it should also explain recursive functional mappings such as:

- **Nucleus → Central Nervous System** (mapping of control functions)
- **Cell Membrane → Skin → Social Boundaries** (mapping of boundary functions)
- **Capillaries → River Systems → Transportation Networks** (mapping of transport functions)

Accordingly, this law concerns not the amplification of **shape**, but the recursive mapping of **structural function** across progressively higher organizational levels.

Section 6. Structural Self-Similarity Law — Why Does Life Continuously Repeat Itself?

Abstract:

Previous sections proposed five fundamental laws of life from the perspective of the Structural Periodic Table: the Law of Structural Periodicity, the Law of Structural Recursion, the Law of Triadic Functional Complementarity, the Law of Structural Hierarchical Transition, and the Law of Structural Mapping. Together they explain how life emerges, develops, and expands across successive organizational levels.

This section proposes a sixth law: **the Law of Structural Self-Similarity**.

Although living systems differ enormously in size, function, and complexity, they repeatedly exhibit analogous structural organization across different hierarchical scales. From genes to cells, from organs to organisms, and from ecosystems to civilizations, similar structural roles continue to reappear.

From the viewpoint of the Structural Periodic Table, what repeats throughout nature is not necessarily identical material, but identical structural organization. This recurring self-similarity provides a possible explanation for why life continuously reproduces itself throughout evolution.

I. Why Does Life Continuously Repeat Similar Structures?

One of the most remarkable characteristics of life is repetition.

Throughout biological evolution, similar organizational patterns repeatedly emerge at different scales. For example:

- genes are composed of repeated nucleotide sequences;
- cells share nearly identical structural organization;
- tissues consist of repeating cellular units;
- organs are assemblies of repeating tissues;
- organisms consist of repeating organ systems.

Although their sizes differ dramatically, their organizational principles remain strikingly similar. Therefore, repetition appears to be a structural phenomenon rather than merely a biological one.

II. Structural Self-Similarity Across Hierarchical Levels

From the perspective of the Structural Periodic Table, life develops through successive structural amplification: **DNA → Cell → Tissue → Organ → Organism**

At each hierarchical level, the three fundamental structural roles reappear:

- **Point (•)** — local generation and information;
- **Line (|)** — connection, transport, and evolution;
- **Circle (O)** — systemic integration and stable closure.

Although each level possesses different biological functions, their structural relationships remain analogous.

Thus, hierarchy does not eliminate structural identity; instead, it preserves it through continuous reorganization.

III. Examples of Structural Self-Similarity

Structural self-similarity can be observed throughout nature.

3.1 Mathematics

Mathematical systems repeatedly generate new structures through recursive expansion. For example: **algebra; geometry; probability theory**.

Although these branches study different objects, each develops from the same underlying structural principles.

3.2 Physics

The three major physical theories also emphasize different structural roles:

- quantum theory emphasizes generation;
- electromagnetism emphasizes propagation;
- gravitation emphasizes structural closure.

Together they reconstruct the same triadic organization.

3.3 Chemistry

Chemical organization similarly exhibits structural repetition: **Elements → Reactions → Molecules**

Simple building blocks continually generate increasingly complex structures while preserving the same organizational logic.

3.4 Biology

Biological systems display structural self-similarity at multiple levels: DNA → Gene → Chromosome → Cell → Tissue → Organ → Organism

Each successive level reorganizes rather than replaces previous structural roles.

IV. Structural Self-Similarity in Living Systems

Many biological structures demonstrate remarkable self-similarity. Examples include:

- branching trees;
- vascular systems;
- bronchial networks;
- neuronal dendrites;
- blood vessels.

Despite serving different biological functions, they all exhibit comparable branching architectures.

Likewise, DNA information expands into complete organs while preserving its structural organization.

The male reproductive system represents a large-scale amplification of sperm-producing structures, whereas the female reproductive system represents the corresponding amplification of ovum-producing structures.

These examples suggest that biological growth frequently enlarges existing structural patterns instead of inventing entirely new ones.

V. Structural Self-Similarity Beyond Biology

The same phenomenon appears beyond living organisms.

Human knowledge also develops through repeated structural organization. For example:

Discipline	Repeated Structural Organization
Mathematics	Algebra – Geometry – Probability
Physics	Quantum – Electromagnetism – Gravitation
Chemistry	Elements – Reactions – Molecules
Life Sciences	Information – Metabolism – Cells

Table 2-11-6-1: Repeated Structural Organization

Although each discipline investigates different phenomena, all exhibit comparable structural organization.

Therefore, structural self-similarity may extend beyond biology into the organization of scientific knowledge itself.

VI. Structural Self-Similarity and the Hexagonal Framework

The Structural Periodic Table proposes that structural repetition continues beyond the natural sciences.

As complexity increases, similar organizational principles may reappear in: **animals; plants; astronomy; politics; economics; psychology.**

These six domains form the basis of the **Hexagonal Structure Framework**, in which the same structural logic is expressed through different material, biological, and social systems.

Thus, structural self-similarity provides an important theoretical bridge between the Structural Periodic Table and the Hexagonal Structure.

VII. Definition of the Law of Structural Self-Similarity

This chapter proposes the following principle: **Structures generated through the triadic framework (Point–Line–Circle) tend to reproduce analogous organizational patterns across different hierarchical scales. Although their size, material composition, and biological function may differ, their structural roles remain fundamentally self-similar.**

Accordingly, life does not continuously invent completely new structures. Instead, it continually reorganizes and amplifies existing structural principles.

VIII. Conclusion

The six structural laws proposed in this chapter together explain why life is capable of continuous emergence, amplification, and evolution.

Among them, the **Law of Structural Self-Similarity** explains why structural organization repeatedly appears throughout different hierarchical levels.

From atoms to organisms, and from organisms to civilizations, structural roles continue to recur.

Therefore, what evolves is not merely biological form, but structural organization itself.

Closing Statement: **Scale changes; structure repeats. Matter evolves, but structural organization remains remarkably**

self-similar. The continuity of life is therefore not merely the continuity of genes, but the continuity of structure itself.

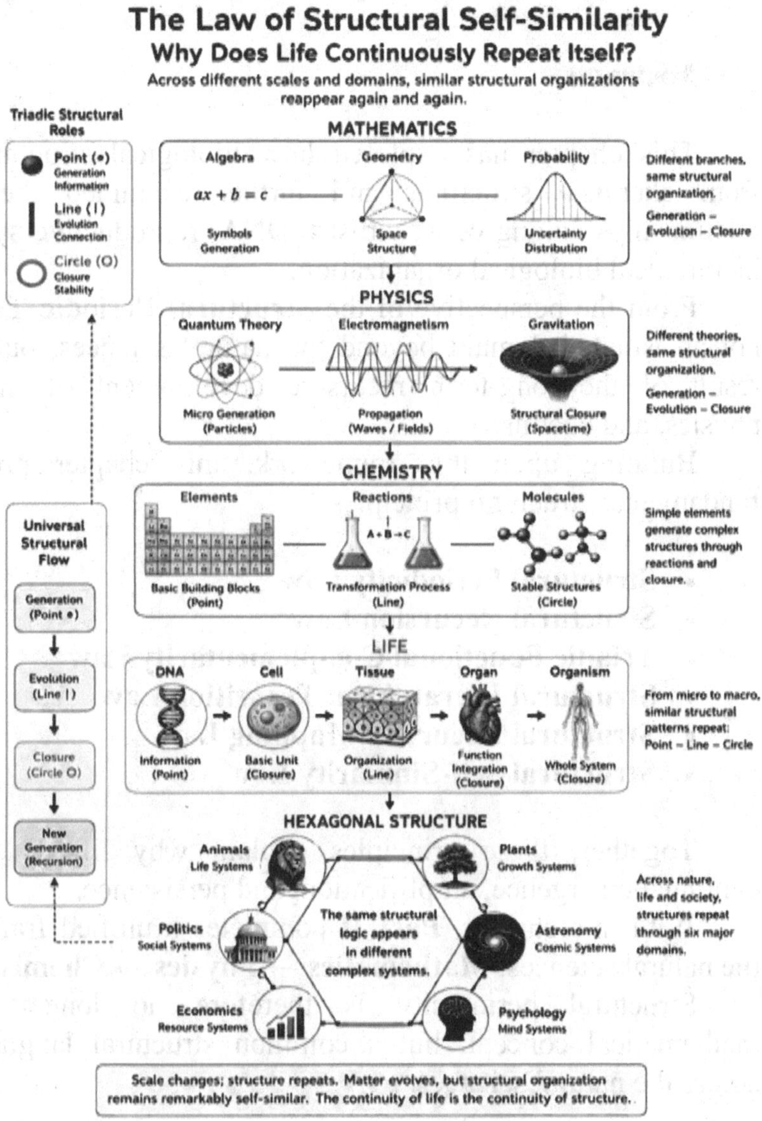

Figure 2-11-6-1: Structural Self-Similarity Law: Why Does Life Continuously Repeat Itself?

Section 7: Chapter Summary – From Chemical Structures to the Structural Periodicity of Life

Abstract:

This chapter has explored how biological structures emerge from chemical structures and further examined the structural relationships among cells, viruses, DNA, reproductive systems, and hierarchical biological organization.

From the perspective of the **Structural Periodic Table**, life is not an isolated domain beyond the natural sciences, but rather the result of the long-term recursive development of mathematics, physics, and chemistry.

Building upon this framework, this chapter proposes six fundamental structural principles:

- **Structural Periodicity Law**
- **Structural Recursion Law**
- **Triadic Functional Complementarity Law**
- **Structural Hierarchical Transition Law**
- **Structural Recursive Mapping Law**
- **Structural Self-Similarity Law**

Together, these principles explain why life is capable of continual emergence, amplification, and persistence.

With this chapter, **Part II** completes a unified framework for the natural sciences: **Mathematics → Physics → Chemistry → Life**

Structural periodicity is therefore no longer merely a mathematical concept, but a common structural language shared across the natural sciences.

I. Life Did Not Appear Independently

Traditional disciplines generally regard mathematics, physics, chemistry, and biology as separate fields.

The **Structural Periodic Table**, however, proposes that life does not introduce entirely new principles. Rather, it represents a new hierarchical level emerging through the continual recursion of existing structural laws.

Thus, life develops through the sequence: **Mathematics → Physics → Chemistry → Life**

This progression represents a continuous unfolding of the natural sciences.

II. Mathematics Provides the Logic of Structure

This book first examined several fundamental mathematical structures:

- recursion,
- groups,
- periodic functions,
- moduli spaces,
- structural spectra.

Together, these demonstrate that nature first possesses an underlying logical architecture.

Accordingly, mathematics answers the question: **Why can structures exist?**

III. Physics Provides Natural Dynamics

Physics subsequently explains how structures evolve through natural processes. Its major structural manifestations include:

- particle spectra,
- field propagation,
- conservation laws,
- phase transitions,
- cosmic hierarchy.

Thus, physics answers: **How do structures evolve?**

IV. Chemistry Provides Material Organization

Chemistry further establishes material organization through:

- the periodic table of elements,
- electron shell structures,
- chemical bonding,
- reaction networks,
- molecular complexity.

Accordingly, chemistry answers: **How do structures become stable forms of matter?**

V. Life Completes Natural Structural Closure

Life extends this progression through: **DNA, cells, tissues, organs, reproductive systems.**

Consequently, **Information → Function → Life → Information**, forms another complete structural cycle.

Life therefore answers: **How can structures sustain themselves?**

VI. A Unified Structural Chain Across the Natural Sciences

The four disciplines may therefore be summarized as follows:

Discipline	Core Question	Structural Role
Mathematics	Why can structures exist?	Logical Foundation
Physics	How do structures evolve?	Natural Dynamics
Chemistry	How do structures stabilize?	Material Organization

Life Science	How do structures continue?	Biological Continuity

Table 2-11-7-1: **Unified Representation of the Four Natural Sciences**

This leads naturally to the sequence: **Mathematics → Physics → Chemistry → Life**
forming a continuous recursive system throughout the natural sciences.

VII. The Six Fundamental Principles of the Structural Periodic Table

Throughout **Part II**, this book progressively establishes six fundamental structural principles.

First: Structural Periodicity Law, Why do stable structures repeatedly appear?

Second: Structural Recursion Law, Why can structures continually expand into higher levels?

Third: Triadic Functional Complementarity Law, Why must life consist of the integrated functions of **Point •, Line ｜, and Circle O**?

Fourth: Structural Hierarchical Transition Law, Why are living systems capable of continually entering higher organizational levels?

Fifth: Structural Recursive Mapping Law, How can genetic information be progressively amplified into complete living organisms?

Sixth: Structural Self-Similarity Law, Why do similar structural patterns repeatedly emerge across different scales and scientific disciplines?

Together, these six principles constitute the proposed theory of **Structural Periodicity in Life**.

More importantly, they themselves form a complete recursive cycle:

Periodicity → Recursion → Complementarity → Hierarchical Transition → Recursive Mapping → Self-Similarity → New Periodicity

This perfectly reflects the central idea of the **Structural Periodic Table**.

Rather than six isolated principles, they constitute a single interconnected structural system. This theoretical framework is not limited to explaining life alone.

In the subsequent volume, **Hexagonal Structure**, these same principles will naturally extend to higher-level systems including: **animals, plants, psychology, politics, economics, astronomy**.

VIII. Unified Expression of the Natural Sciences

The entire second part may therefore be summarized as follows:

Discipline	**Point •**	**Line ∣**	**Circle O**
Mathematics	Initial Rules	Recursive Expansion	Structural Closure
Physics	Local Sources	Dynamic Propagation	Stable Structures
Chemistry	Elements	Reaction Pathways	Molecular Structures
Life Science	Genetic Information	Functional Execution	Living Systems

Table 2-11-7-2. Unified Structural Periodicity Across the Natural Sciences

Although the objects of study continually change, **Point → Line → Circle**, remains fundamentally unchanged.

The triadic structural framework therefore expands naturally into the **Hexagonal Structure**, giving rise to higher-order systems such as: animals, plants, psychology, society, civilization, astronomy.

Once life emerges, structural evolution does not terminate.

What ultimately persists is not any particular scientific theory, but **structure itself**.

Accordingly, the present volume, **Structural Periodic Table**, establishes the unification of the natural sciences.

Its companion volume, **Hexagonal Structure**, will continue this framework by investigating how life, society, civilization, and the universe develop into higher-order structural periodic systems, thereby extending the theory toward the unification of life and civilization.

Structural Conclusion: Mathematics reveals the logical foundations of structure. Physics realizes the dynamics of structure. Chemistry stabilizes material structure. Life perpetuates structural evolution. Together, these four disciplines constitute the unified natural-science framework of the **Structural Periodic Table**.

References (Chapter 11)

Cell Biology and Molecular Biology

[1] **Alberts, B., Johnson, A., Lewis, J., Morgan, D., Raff, M., Roberts, K., & Walter, P.** (2022). *Molecular Biology of the Cell* (7th ed.). Garland Science.

[2] **Lodish, H., Berk, A., Kaiser, C. A., et al.** (2021). *Molecular Cell Biology* (9th ed.). W. H. Freeman.

[3] **Campbell, N. A., Urry, L. A., Cain, M. L., et al.** (2020). *Campbell Biology* (12th ed.). Pearson.

DNA, RNA, and Genetics

[4] **Watson, J. D., Baker, T. A., Bell, S. P., Gann, A., Levine, M., & Losick, R.** (2014). *Molecular Biology of the Gene* (7th ed.). Pearson.

[5] **Watson, J. D., & Crick, F. H. C.** (1953). "Molecular Structure of Nucleic Acids." *Nature*, **171**, 737–738.

[6] **Crick, F. H. C.** (1970). "Central Dogma of Molecular Biology." *Nature*, **227**, 561–563.

Viruses and Evolution

[7] Flint, S. J., Enquist, L. W., Racaniello, V. R., & Skalka, A. M. (2020). *Principles of Virology* (5th ed.). ASM Press.

[8] Koonin, E. V. (2011). *The Logic of Chance: The Nature and Origin of Biological Evolution*. FT Press.

Systems Biology and Complexity

[9] Kitano, H. (2002). "Systems Biology: A Brief Overview." *Science*, **295**, 1662–1664.

[10] Barabási, A.-L. (2016). *Network Science*. Cambridge University Press.

[11] Mitchell, M. (2009). *Complexity: A Guided Tour*. Oxford University Press.

[12] Simon, H. A. (1962). "The Architecture of Complexity." *Proceedings of the American Philosophical Society*, **106**(6), 467–482.

Structural Theory

[13] Weyl, H. (1952). *Symmetry*. Princeton University Press.

[14] Thompson, D. W. (1917/1992). *On Growth and Form*. Cambridge University Press.

Structural Periodicity Theory

[15] Chang, J. (2013). *Grand Ultimate Theory*. Universal Publishing.

[16] Chang, J. (2026). *Recursive Self-Organizing Evolutionary Breakthrough: Unifying Natural, Social and Life Sciences*. Universal Publishing.

[17] Chang, J. (2026). *Metastructural Unification (Physics Volume): Unifying Electromagnetism, Gravitation and Quantum Theory*. Universal Publishing.

[18] Erwin Schrödinger (1944). *What Is Life?* Cambridge University Press.

Part III - Methodology of the Structural Periodic Table : Language, Prediction, and Boundaries

Figure 3: A crop circle illustrating Galactic Civilization. For details, see book 《Crop Circle》.

Structural Theme:

Every mature scientific theory ultimately requires not only explanatory power, but also a methodology. A methodology defines the language through which a theory describes reality, the principles by which it makes predictions, and the boundaries beyond which it must be revised or abandoned.

The final part of this book therefore discusses the methodology of the Structural Periodic Table through three perspectives: Language, Prediction, and Boundary.

The methodological framework of the **Structural Periodic Table** follows a continuous sequence: **Generation → Stability → Periodicity → Structural Limit**

This sequence forms the fundamental methodology underlying the entire book. The final three chapters complete the theoretical framework through another progression: **Language → Prediction → Boundary**, together, these form a complete logical cycle.

The first ten chapters establish the theoretical foundations. Chapter 11 introduces the fundamental structural laws. Chapters 12–14 then explore the language of the theory, its predictive capability, and its theoretical boundaries.

Consequently, the overall structure of this book may be summarized as: **Foundation → Verification → Adjudication → Prediction → Boundary**

This progression represents the complete developmental path of a scientific theory — from its initial formulation, through empirical and structural evaluation, to its predictive power and ultimate limitations.

Chapter Twelve: The Unified Structural Language of the Natural Sciences

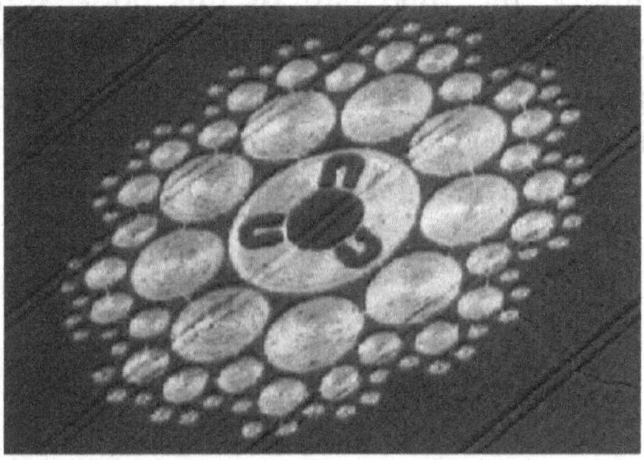

Figure 3-12: A crop circle illustrating Galactic Civilization. For details, see book 《Crop Circle》.

Chapter Overview:

The triadic structures identified throughout mathematics, physics, and chemistry correspond to the three fundamental structural roles: **Line | ↔ Circle O ↔ Point •**

Together, they constitute the common foundation of structural periodicity across the natural sciences.

Throughout the history of science, mature theories have often concluded by examining their own methodological foundations.

Isaac Newton ultimately turned to questions of scientific method.

Albert Einstein devoted his later work to the search for a unified theory.

Modern mathematics culminates in the study of axiomatic systems and the foundations of formal reasoning.

Likewise, this book concludes by examining three fundamental questions: **Language → Prediction → Boundary**

This progression is consistent with the natural evolution of scientific theories.

A mature theory must not only explain existing phenomena, but also establish a coherent language, generate testable predictions, and clearly define the limits of its own applicability.

Accordingly, the final part of this book shifts its focus from constructing the **Structural Periodic Table** to evaluating its scientific significance as a unified structural language for the natural sciences. It explores how structural language may serve as a framework for explanation, prediction, and theoretical self-assessment.

Section 1. The Unified Structural Periodicity of Mathematics, Physics, and Chemistry

Abstract:

Part II of this book examined structural periodicity from the perspectives of mathematics, physics, and chemistry. Although these three disciplines investigate different objects, the Structural Periodic Table reveals that they share a remarkably similar underlying organization.

Mathematics describes the logical generation of abstract structures; physics explains the dynamic evolution of natural systems; chemistry demonstrates the material organization of matter. Viewed through the lens of structural periodicity, these disciplines are not isolated domains but successive manifestations of a common structural process.

This chapter introduces a unified structural language based on three fundamental structural roles—**Point •, Line ｜, and Circle O**—to describe the common mechanism underlying the natural sciences. It argues that complex structures emerge through the continual interaction of **generation, evolution, and structural closure**, thereby providing a unified framework for interdisciplinary structural research.

I. Introduction

1.1 Structural Periodicity in Mathematics

Within mathematics, structural periodicity appears as the generation and stabilization of abstract structures.

Recursive structures, group theory and symmetry, periodic functions, moduli spaces, and structural spectra collectively illustrate the fundamental developmental pathway of mathematical structures: **Recursive Generation → Symmetry Constraint → Periodic Repetition → Moduli Classification → Structural Spectrum**

This progression reveals a recurring relationship between structural generation and structural stability within mathematical systems.

1.2 Structural Periodicity in Physics

Within physics, structural periodicity appears as the emergence of stable hierarchies in natural dynamical systems.

Particle spectra, field modes, conservation laws, phase transitions, and cosmic hierarchies together represent the typical evolutionary pathway of physical structures: **Particle Spectrum → Field Modes → Conservation Constraints → Phase Transition → Cosmic Hierarchy**

This progression demonstrates that physical structures originate from localized sources and subsequently evolve into stable configurations through propagation and constraint.

1.3 Structural Periodicity in Chemistry

Within chemistry, structural periodicity appears as the hierarchical organization of material structures.

The periodic table of elements, electron shell structures, recurring bonding patterns, chemical reaction networks, and molecular complexity collectively reveal the characteristic developmental pathway of matter: **Periodic Elements → Shell Constraints → Bond Formation → Reaction Network Evolution → Molecular Complexity**

This sequence illustrates the hierarchical expansion of material organization from individual elements to highly complex molecular systems.

II. Comparison of the Three Systems

A comparison of these three disciplines reveals that, despite their different subjects of study, they all exhibit remarkably similar structural patterns.

To describe this common organization, this book introduces a unified triadic structural language:

- **Point •** — Structural Generation Source
- **Line** | — Structural Evolution Pathway
- **Circle O** — Structural Closure and Stability

Within this framework, mathematics, physics, and chemistry correspond as follows:

Discipline	**Point • (Genera-tion)**	**Line | (Evolution)**	**Circle O (Stability)**	**Periodic Manifestation**
Mathe-matics	Initial Rules	Recursive Expansion	Structural Closure	Structural Spectrum
Physics	Local Sources	Dynamic Propagation	Stable Modes	Particle Spectrum
Chemistry	Elements	Reaction Pathways	Molecular Structures	Periodic Table

Table 3-12-1-1: Triadic Structural Correspondence Across Mathematics, Physics, and Chemistry

This correspondence suggests that structural periodicity is not unique to any individual discipline, but represents a universal structural mechanism throughout the natural sciences.

More generally, the three structural roles form a continuous recursive cycle: **Point → Line → Circle → New Point**

This cycle implies that:

1) Local structural sources initiate new structures.
2) Structures expand through evolutionary pathways.
3) Stable closure emerges under structural constraints.
4) Stable structures become the foundation for higher-level generation.

Accordingly, the complexity observed throughout nature may be interpreted as repeated manifestations of this structural cycle across different scales.

• In mathematics, the cycle appears as the generation and classification of abstract structures.

• In physics, it appears as propagation and stabilization within dynamical systems.

• In chemistry, it appears as material formation and reaction-driven organization.

A cross-disciplinary conclusion therefore emerges: Structural periodicity is not an isolated phenomenon within individual sciences, but a universal expression of the recurring relationship among **generation, evolution, and stability**.

Consequently, mathematics, physics, and chemistry should not be regarded as independent disciplines, but rather as successive hierarchical stages: **Mathematical Structure → Physical Structure → Chemical Structure**

This hierarchy suggests that the complexity of the natural sciences can be described using a common structural language.

III. A Unified Structural Framework for Mathematics, Physics, and Chemistry

The triadic structural framework provides a unified methodology for interdisciplinary structural research.

It also expresses the central proposition of this entire book: **Point •, Line ｜, and Circle O are the three fundamental drivers of structure.**

Discipline	Structural Prototype	Triadic Interpretation	Summary
Mathematics	Chapter 5, Section 1	Chapter 5, Section 2	Chapter 5, Section 3
Physics	Chapter 7, Section 1	Chapter 7, Section 2	Chapter 7, Section 3

Chemistry	Chapter 9, Section 1	Chapter 9, Section 2	Chapter 9, Section 3

Table 3-12-1-2: Unified Structural Periodicity Across the Natural Sciences

At this point, the overall framework of Part II becomes fully integrated, forming the following logical progression:

Mathematics (Structural Language) → Physics (Natural Dynamics) → Chemistry (Material Organization) → Triadic Structural Unification

---

IV. The Triadic Structural Language: Point •, Line ｜, and Circle O

By comparing structural periodicity across mathematics, physics, and chemistry, a remarkable structural isomorphism becomes apparent.

Although these disciplines investigate different objects, they all exhibit the same fundamental cycle of **generation, evolution, and stability**.

To unify this recurring structural pattern, the Structural Periodic Table introduces the following triadic language:

- **Point •** — Structural Generation Source
- **Line ｜** — Structural Evolution Pathway
- **Circle O** — Structural Closure and Stability

From the perspective of structural periodicity, these three disciplines naturally unfold as a continuous hierarchy:

Mathematical Structure → Physical Structure → Chemical Structure

Within this hierarchy:

- Mathematics provides the abstract language of structure.
- Physics describes the dynamics of natural systems.
- Chemistry demonstrates the stable organization of matter.

Accordingly, complex structures in nature do not arise randomly; they emerge through the continual interaction of generation, evolution, and stabilization.

This structural cycle may therefore be summarized as: **Point → Line → Circle → New Point,**

namely:

1) Local structural sources initiate new structures.

2) Structures evolve through continuous pathways.

3) Stable structures emerge under constraint.

4) Stable structures become the foundation for subsequent hierarchical development.

This recursive cycle continually repeats across different scales, giving rise to the hierarchical organization observed throughout the natural sciences.

V. General Structural Periodicity of the Natural Sciences

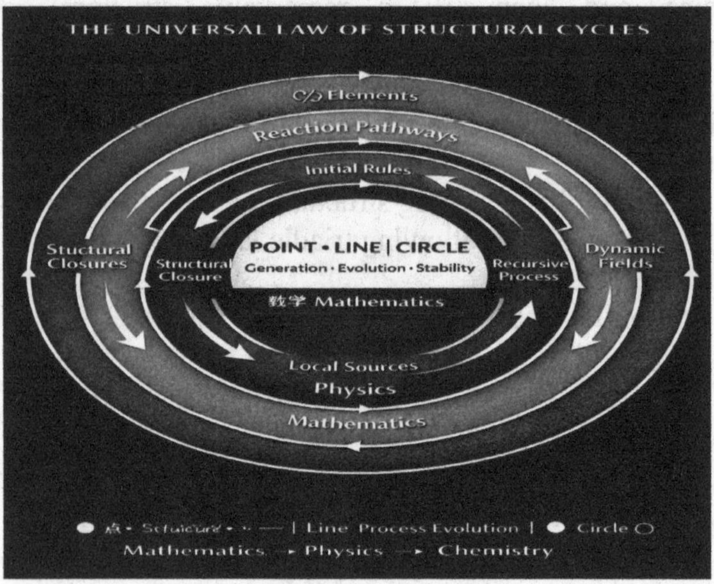

Figure 3-12-1-1: General Structural Periodicity Framework of the Universal Law *(A three-layer concentric structural diagram.)*

VI. Structural Manifesto

Viewed from a broader perspective, natural systems reveal a common structural principle.

The generation of abstract structures in mathematics, the propagation of dynamic processes in physics, and the organization of matter in chemistry all follow the same fundamental structural cycle.

This universal cycle may be expressed using three elementary structural roles:

- **Point** • generates structure.
- **Line** | drives structural evolution.
- **Circle O** establishes structural stability.

Accordingly, periodic structures throughout the natural sciences are not isolated phenomena but manifestations of the continuous interaction among these three fundamental structural dynamics.

Point •, Line |, and Circle O together constitute the common structural foundation of periodicity throughout the natural sciences.

Section 2. Ten Structural Questions – From the Periodic Table of Elements to the Structural Periodic Table

Abstract:

Beginning with the periodic table of elements, this book seeks to establish a more general framework of **structural periodicity**.

Throughout this investigation, a series of fundamental questions continually emerge:

- What is periodicity?
- Why do structures repeat?
- Why do hierarchical levels arise?
- Why do stable structural windows exist?
- Why do structural limits appear?

Although these questions seem diverse, they ultimately converge upon a single underlying issue: **How are structures generated, stabilized, and renewed?**

This section condenses the methodology developed throughout the book into **ten fundamental questions** and proposes a unified structural interpretation of each.

I. Ten Fundamental Questions

1.1 Question One: What Is Periodicity?

Periodicity is not merely the repetition of numerical values.

Within the Structural Periodic Table, periodicity is better understood as the recurring emergence of **stable structural windows** generated through structural recursion.

Therefore, **Periodicity ≠ Repetition**

Periodicity = The Recurring Emergence of Stable Structures

1.2 Question Two: What Is the Deepest Secret of the Periodic Table?

The true essence of the periodic table does not lie solely in electron shells, quantum numbers, or symmetry.

Its deeper significance may be summarized as follows: **Finite structures achieve stable closure under constraints, and each completed closure initiates a new hierarchical level.**

Consequently, **Structural closure and structural renewal are more fundamental than periodicity itself.**

1.3 Question Three: What Is Structural Closure?

Structural closure should not be regarded as an endpoint. Instead, it signifies that:

- local generation achieves global stability;
- global stability becomes the foundation for the emergence of a higher structural level.

Thus, **Every closure is simultaneously an ending and a new beginning.**

1.4 Question Four: What Is the Relationship Between Generation, Stability, and Renewal?

Throughout nature, one repeatedly observes the cycle: **Generation → Stability → Renewal**

This pattern appears in:

- chemical elements,
- stellar evolution,
- biological evolution,
- social development.

Accordingly, **Periodicity is the recurring consequence of generation, stability, and renewal.**

1.5 Question Five: Which Comes First—Point, Line, or Circle?

Strictly speaking, none of the three possesses absolute priority.

Point •, Line ｜, and Circle O always coexist.

Their only difference lies in which structural role dominates at a given stage.

Therefore, **Structural periodicity does not arise because the three appear sequentially, but because their relative dominance continually changes.**

1.6 Question Six: Why Do Structural Transitions Occur?

Hierarchical transitions are never triggered by a single factor. Instead, they emerge from the imbalance among:

- local structural tension,
- evolutionary direction,
- global structural constraints.

Hence, **Structural transitions occur when structural tension reaches a critical window.**

1.7 Question Seven: What Is the Essence of Periodic Structures?

Periodicity is not simple repetition. Its deeper nature is better understood as: **Structural transitions between successive stable windows.** Accordingly,

- elemental periodicity,
- biological rhythms,
- economic cycles,
- civilizational transformations

may all be interpreted as structural oscillations occurring at different scales.

1.8 Question Eight: What Are the Fundamental Variables of Structural Periodicity?

Ultimately, this book reduces structural complexity to three fundamental variables:

- **Point •** — Local structural tension and perturbation
- **Line |** — Propagation and structural evolution
- **Circle O** — Constraint and stable structural windows

Together, these constitute the minimal expression of structural periodicity.

1.9 Question Nine: Which Variable Is Irreducible?

Viewed individually, Point cannot be eliminated. Line cannot be eliminated. Circle cannot be eliminated.

Yet what is truly irreducible is not any single variable, but rather **the coupling among all three.**

Therefore, **Structural periodicity arises from triadic interaction rather than from any single driving factor.**

1.10 Question Ten: What Governs Structural Renewal?

Renewal is not caused solely by:

- an explosive Point,
- the collapse of Circle,
- or the instability of Line.

Instead, renewal occurs when **Point, Line, and Circle simultaneously approach their critical states.**

Accordingly, **Periodic renewal is fundamentally a phenolmenon of triadic structural resonance.**

II. Unifying the Ten Questions

The preceding discussion allows structural periodicity to be condensed into its simplest conceptual framework:

- **Point •** — Local Structural Tension
- **Line |** — Propagation and Evolution
- **Circle O** — Constraint and Structural Closure

Together they generate the recurring structural cycle: **Generation → Stability → Renewal**

Thus, structural periodicity is not simple repetition, but the continual unfolding of **Point → Line → Circle → New Point,** across multiple hierarchical scales.

III. Structural Periodicity Manifesto

The periodic table of elements reveals the periodic organization of the material world.

The **Structural Periodic Table** seeks to reveal the structural organization underlying complex systems.

As system complexity increases, strict periodicity gradually weakens, while structural repetition becomes increasingly dominant.

Accordingly, the periodic table predicts **unknown elements**; the Structural Periodic Table seeks to identify **unknown structures**.

From the periodicity of the elements to the boundary of life, and from the natural sciences to complex systems, the true subject of structural periodicity is **not periodicity itself**, but **the structural mechanisms that generate periodicity**.

Structural Conclusion:

The Periodic Table explains why matter repeats.

The Structural Periodic Table seeks to explain why structure itself repeats.

The recurring emergence of stable structures — not simple repetition — is the deepest language shared by the natural sciences.

Section 3. Point •, Line | , and Circle O – The Three Fundamental Structural Roles

Abstract:

This book proposes that **Point •, Line | , and Circle O** are not merely geometric symbols, nor simply classificatory tools used in mathematics, physics, or chemistry. Rather, they represent the **Three Fundamental Structural Roles** required by every stable system of existence. Within this framework:

• **Point •** corresponds to **Generation**, representing the emergence of new localized existence.
• **Line |** corresponds to **Evolution**, representing direction, propagation, interaction, and structural connection.
• **Circle O** corresponds to **Closure**, representing stability, self-maintenance, and the formation of an integrated whole.

Any system capable of sustained existence — whether physical, biological, or social — must undergo these three fundamental processes: **generation, evolution, and closure**.

Without any one of these roles, no structure can remain stable. Accordingly, this book argues that **Point •, Line | , and Circle O** are not arbitrary classifications, but constitute the **Minimum Structural Language of Existence**.

I. Why Are Three Structural Roles Necessary?

When systems are examined across different scales, one common phenomenon repeatedly emerges.

Stable existence never appears instantaneously. Instead, every stable system passes through three successive stages: **The emergence of local structures → the establishment of structural connections → the formation of stable closure.**

Whether the subject is mathematical structures, physical systems, chemical reactions, or living organisms, these three stages continually recur.

Therefore, they are not discipline-specific principles but universal principles governing existence itself.

II. The First Role: Point • — Generation

Point • is not merely a geometric point. It represents the emergence of a new localized existence.

It possesses three essential characteristics:

- **Irreducibility**
- **Identity**
- **Generativity**

Every new structure must first establish an independent local unit before further organization becomes possible.

Accordingly, Point • is not an endpoint but the beginning of all structural evolution. It answers the fundamental question: **What can begin to exist?**

III. The Second Role: Line | — Evolution

Once local structures have emerged, they no longer remain isolated. Instead, relationships begin to form through:

- interaction,
- information transfer,
- energy propagation,
- pathway formation.

Accordingly, Line | is not simply a geometric line.

It represents the emergence of direction, propagation, and structural evolution.

Without Line |, local entities would remain isolated, and higher-order organization could never develop.

It answers the question: **How do structures develop?**

IV. The Third Role: Circle O – Closure

If evolution continued indefinitely without organization, no stable system could ever exist. Every long-lasting system must therefore eventually achieve structural closure. Closure signifies:

- internal coordination,
- boundary formation,
- functional stability,
- sustainable existence.

Accordingly, Circle O represents the stable whole formed through structural completion.

It answers the question: **Why can structures persist?**

V. Why Does the Entire Universe Require These Three Roles?

This book proposes a fundamental principle: **Stable existence is not an object, but a process.**

That process necessarily contains three indispensable structural roles: **Generation → Evolution → Closure**

- Without generation, no new structures appear.
- Without evolution, structures cannot develop.
- Without closure, systems cannot remain stable.

Therefore, every enduring system must simultaneously possess:

- generation,
- evolution,
- closure.

This is not merely a law of any particular scientific discipline.

It represents the **minimum organizational principle of existence itself**.

VI. The Three Roles Form the Minimum Structural Cycle

The three structural roles do not exist independently. Instead, they form a continuous recursive cycle: **Generation → Evolution → Closure → New Generation**

or, in the symbolic language of this book: **Point • → Line | → Circle O → New Point •**

Whenever one structural cycle reaches closure, new localized structures emerge, initiating another cycle of evolution.

Accordingly, the universe is not composed of static arrangements, but of continuous recursive processes of **generation, evolution, closure, and regeneration.**

VII. Point •, Line |, and Circle O Are Roles, Not Categories

Traditional science generally classifies objects according to their physical nature:

- particles,
- fields,
- molecules,
- organisms,
- societies.

This book proposes a deeper form of classification.

What truly matters is not **what an object is**, but **which structural role it performs**. The same object may perform different structural roles at different hierarchical levels.

Conversely, entirely different objects may perform the same structural role. Therefore, the truly universal entities are not particular objects, but **Structural Roles**.

VIII. Toward a Three-Dimensional Structural Framework

The theoretical framework may now be organized into three complementary dimensions.

Historically, the progression developed as: **Line → Circle → Point**

It subsequently expanded into: **Mathematics → Physics → Chemistry → Life**

It may now be further generalized as: **Micro-scale → Meso-scale → Macro-scale**

These three directions together form a three-dimensional structural coordinate system. For example:

- **Scale Axis:** Atoms → Molecules → DNA → Cells → Organs → Organisms → Planets → Galaxies → Cosmic Structures
- **Structural Role Axis:** Point • → Line ｜ → Circle O
- **Disciplinary Axis:** Mathematics → Physics → Chemistry → Life → Society → Civilization

Accordingly, the Structural Periodic Table begins to reveal a genuine **three-dimensional structural coordinate system**, within which scale, function, and discipline become unified through a common structural language.

IX. Summary

This book ultimately proposes the following fundamental interpretation:

- **Point • is the Generator.**
- **Line ｜ is the Evolver.**
- **Circle O is the Stabilizer.**

Together they constitute the minimum structural language required by every stable system of existence.

Consequently, Point •, Line ｜, and Circle O should no longer be regarded merely as classificatory symbols, but as the **fundamental structural roles of ontology**.

Structural Conclusion: The universe is not composed merely of objects, but of **structural roles**. **Generation creates existence. Evolution connects existence. Closure sustains existence.**

Accordingly, **Point •, Line ｜, and Circle O** are not simply geometric symbols, but the three fundamental structural roles required by every stable system in the universe.

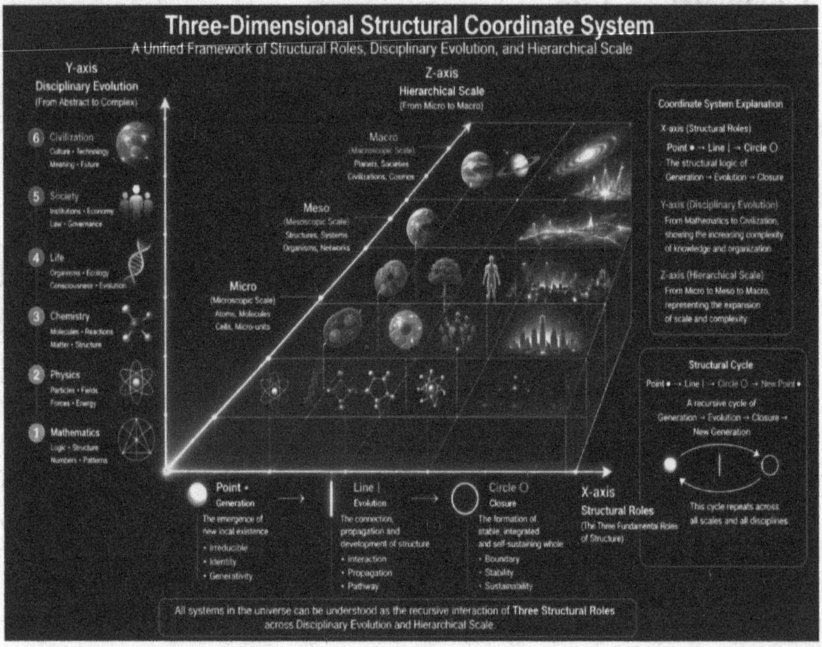

Figure 3-12-3-1: Three-Dimensional Structural Coordinate System

Note: The three coordinate axes can be designed as follows:

- **X-axis:** Point • → Line ｜ → Circle O（Structural Roles）
- **Y-axis:** Mathematics → Physics → Chemistry → Life → Society → Civilization（Disciplinary Evolution）
- **Z-axis:** Micro → Meso → Macro（Hierarchical Scale）

This diagram is highly likely to become one of the most representative overview illustrations in the entire 《Structural Periodic Table》 book, and it also serves as a natural bridge to the next volume, 《Hexagonal Structures》.

Section 4. From the Language of Classification to the Language of Structural Dynamics

Abstract:

The history of modern science is not merely the accumulation of knowledge, but also the continuous evolution of scientific language.

This book proposes that scientific language has progressed through at least three major stages:

- **The First Generation:** Classification, centered on identifying and organizing objects.
- **The Second Generation:** Generation, centered on explaining how systems emerge and evolve.
- **The Third Generation:** Structural Dynamics, centered on understanding how structures are generated, evolve, stabilize, and continue across hierarchical levels.

The **Structural Periodic Table** is therefore not simply another classification system.

Rather, it proposes a new scientific language capable of describing the generation, evolution, stabilization, and recursive continuation of structures across different scientific disciplines.

I. The First Generation of Scientific Language: Classification (Point •)

The earliest mission of science was to identify and classify the objects found in nature. Typical examples include:

- Biological taxonomy;
- The Periodic Table of Elements;
- Astronomical classification;
- Mineral classification.

The central question during this stage was: **What exists?**

Classification established the essential foundation of modern science. Among all classification systems, the Periodic Table of Elements remains one of the greatest achievements. It not only organized known elements but also successfully predicted previously undiscovered ones.

Nevertheless, classification possesses an inherent limitation. It describes **what things are**, but cannot fully explain **why they exist** or **how they came into being.**

II. The Second Generation of Scientific Language: Generation (Line |)

With the development of modern mathematics, physics, evolutionary biology, and complexity science, scientific attention gradually shifted from classification toward generation.

Representative examples include:

- Differential equations describing dynamic evolution;
- Evolutionary theory explaining the emergence of life;
- Cosmology explaining the origin of the universe;
- Self-organization theory explaining the emergence of complex systems.

Science therefore began asking a new question: **How does the world arise?**

Scientific language consequently evolved from static classification toward dynamic generation. The primary object of investigation shifted from **objects** to **processes**.

III. The Third Generation of Scientific Language: Structural Dynamics (Circle O)

This book proposes a further step. Generation alone is still insufficient. Every stable system must ultimately undergo three fundamental structural processes:

- **Generation**

- **Evolution**
- **Closure**

Consequently, understanding only the origin of structures cannot fully explain their long-term existence. The deeper questions become:

- How do structures emerge?
- How do structures evolve?
- Why do structures become stable?
- Why do stable structures generate new hierarchical levels?

These questions define the domain of **Structural Dynamics**.

Unlike conventional theories, Structural Dynamics is not concerned with particular objects. Instead, it investigates the universal structural principles shared by all objects.

IV. Why Is a New Scientific Language Necessary?

Each scientific discipline has developed its own mathematical framework. For example:

- Mathematics employs axiomatic systems.
- Physics employs dynamical equations.
- Chemistry employs reaction models.
- Biology employs evolutionary models.

Each framework has achieved remarkable success.

However, they operate largely within their own disciplinary domains. What remains absent is a common language capable of describing the shared structural behavior of mathematics, physics, chemistry, and life science simultaneously.

This book therefore proposes that **Point •, Line |, and Circle O** should not be regarded as merely new geometric symbols. Instead, they constitute a **cross-disciplinary structural language**, representing respectively:

- structural generation;

- structural evolution;
- structural closure.

They are therefore not individual theories, but the universal structural roles underlying many different theories.

V. From Classification to Structural Dynamics

The historical development of scientific language may be summarized in three stages:

Stage One: Classification / answers: **What exists?**
Stage Two: Generation / answers: **How does it arise?**
Stage Three: Structural Dynamics / answers: **Why can stable structures continue to exist?**

Accordingly, the focus of scientific inquiry has gradually shifted: **Objects → Processes → Structural Dynamics**

or, in the symbolic language of this book: **Point • → Line | → Circle O**

VI. The Significance of Structural Language

If the Periodic Table organized the material world, then structural language seeks to organize the world of knowledge itself.

It offers a unified framework capable of reinterpreting mathematics, physics, chemistry, and life science from a common structural perspective.

Moreover, such a language may eventually provide a common conceptual framework through which artificial intelligence can organize and understand scientific knowledge.

Accordingly, the **Structural Periodic Table** introduces not merely another classification, but a fundamentally new mode of scientific expression.

VII. Summary

Scientific progress is driven not only by the discovery of new knowledge, but also by the creation of new scientific languages.

• **Classification** enables science to identify the world.
• **Generation** enables science to explain the emergence of the world.
• **Structural Dynamics** seeks to explain why the world is capable of sustained existence.

Within this framework, **Point •, Line 丨, and Circle O** provide the most concise expression of this evolution in scientific language.

Structural Conclusion:

Every scientific revolution has been accompanied by the emergence of a new scientific language.

Classification describes **what exists**. Generation explains **how things emerge**. Structural Dynamics seeks to explain **why stable structures persist and continue to evolve**.

Accordingly, the framework of **Point •, Line 丨, and Circle O** should be understood not merely as another system of classification, but as a new language for describing structural evolution itself.

Methodological Significance:

The most important contribution of this chapter is not the introduction of additional scientific content, but the elevation of the theoretical perspective.

The **Structural Periodic Table** should therefore not be interpreted simply as a new classification method. Instead, it should be regarded as **A New Scientific Language**.

From this standpoint, traditional concepts such as the **Five Elements, Yin–Yang**, and classical medical theories are no longer treated as evidence supporting this framework. Rather, they become historical knowledge systems that can be **reinterpreted through a unified structural language**.

Accordingly, the progression of this chapter becomes increasingly coherent:

1) Why does the universe require three fundamental structural roles? (Section 3)

2) Why does science require a new structural language? (Section 4)

3) How can this language reinterpret both ancient and modern knowledge systems? (Section 5)

Through this progression, the **Structural Periodic Table** advances beyond individual scientific disciplines and enters the realm of **scientific methodology**, providing a unified structural language for understanding both science and civilization.

Section 5. Point •, Line |, and Circle O – A Modern Interpretation of Ancient Structural Thinking

Abstract:

Throughout the development of human civilization, different cultures have attempted to explain the complexity of the world using the smallest possible set of fundamental concepts.

Chinese civilization developed theories such as **Yin–Yang**, the **Three Powers (Heaven–Earth–Humanity)**, the **Five Elements**, and the holistic principles of Traditional Chinese Medicine.

Western civilization proposed the theory of the **Four Classical Elements** and systems of logical classification.

Indian philosophy likewise developed comprehensive frameworks describing the composition of nature and the order of life.

Although these traditions emerged long before modern science, they share one common aspiration: **to identify the minimum structural principles governing the organization of the universe.**

This book argues that **Point •, Line |, and Circle O** do not replace these traditional systems of thought. Rather, they constitute a **Modern Structural Language** capable of reinterpreting the common characteristics of **Ancient Structural Thinking** within a unified scientific framework.

I. Why Did Ancient Civilizations Seek Minimum Structures?

Ancient civilizations developed before experimental science existed. Consequently, their primary objective was to discover the most fundamental principles capable of explaining the natural world.

Although different cultures adopted different symbolic systems, they all attempted to answer the same fundamental question: **Can the complexity of the universe be explained through a small number of fundamental structural roles?**

Accordingly, Ancient Structural Thinking may be understood as humanity's earliest attempt to construct a universal structural language.

II. Yin–Yang: The Earliest Structural Relation

The concept of **Yin–Yang** emphasizes:

- opposition and complementarity;
- dynamic balance;
- continuous transformation;
- unity through interaction.

Rather than describing particular objects, Yin–Yang expresses the relationships through which stable systems maintain dynamic equilibrium.

Accordingly, Yin–Yang represents **Structural Relations**, rather than material entities.

Figure 3-12-5-1: Galactic civilization crop circle displays Yin-Yang symbol.

III. The Three Powers: A Hierarchical Structural Perspective

The traditional concept of **Heaven–Earth–Humanity** does not merely describe three separate entities. Instead, it expresses the structural unity of three interconnected hierarchical levels.

It emphasizes:

- local existence;
- global organization;
- coordinated interaction.

Accordingly, the Three Powers represent an early form of **Hierarchical Structural Thinking**.

Model	Fundamental Units
Yin–Yang	2
Three Powers (Heaven–Earth–Humanity)	3
Four Images	4
Five Elements	5
Point • Line ｜ Circle O	3

Table 3-12-5-1: Five Traditional Models for Describing the World

---

IV. The Five Elements: A Dynamic Structural Cycle

The **Five Elements** should not be interpreted as five physical substances. Instead, they are better understood as five interacting modes of structural transformation.

Their fundamental principles include:

- mutual generation;
- mutual restraint;
- transformation;
- balance;
- cyclic recurrence.

Accordingly, the Five Elements emphasize **dynamic structural networks**, rather than static categories.

They represent an early understanding of systemic evolution.

Five Elements	**Structural Tendency**
Wood	Growth
Fire	Expansion
Earth	Integration
Metal	Contraction
Water	Descent

Table 3-12-5-2: The Five Dynamic Tendencies of the Five Elements

---

V. Traditional Chinese Medicine: Holistic Structural Thinking

Traditional Chinese Medicine focuses not on isolated organs, but on systemic coordination. Its principal concepts include:

- relationships among the internal organs;
- meridian networks;
- Yin–Yang balance;
- Five-Element correspondences.

These concepts emphasize that local changes ultimately influence global structural stability.

Accordingly, Traditional Chinese Medicine represents a form of **Holistic Structural Thinking**.

Traditional Concept	Structural Periodic Interpretation
Yin–Yang Balance	Circle O — Structural Closure
Five-Element Interactions	Line ∣ — Structural Evolution
Jing–Qi–Shen	Point • — Structural Generation
Meridian System	Line ∣ — Information and Flow Network
Organ Integration	Circle O — Systemic Closure

Table 3-12-5-3: Point–Line–Circle Interpretation of Traditional Chinese Medicine

VI. The Twelve Meridians and the Six Structural Laws

Traditional Chinese Medicine describes the **Twelve Primary Meridians**, which form the principal pathways through which **Qi** and blood circulate throughout the body.

These meridians directly connect the internal organs and are traditionally organized according to the Five Elements. They include:

- Three Yin Meridians of the Hand;
- Three Yang Meridians of the Hand;
- Three Yang Meridians of the Foot;
- Three Yin Meridians of the Foot.

Within the framework proposed in this book, the six structural laws establish the fundamental organizational principles of living systems:

- Structural Periodicity Law;
- Structural Recursion Law;
- Triadic Functional Complementarity Law;
- Structural Hierarchical Transition Law;

- Structural Recursive Mapping Law;
- Structural Self-Similarity Law.

From a structural perspective, the meridian system may be interpreted not as a classification of anatomical entities, but as an integrated network comprising:

- **Information Networks**
- **Flow Networks**
- **Connection Networks**

Accordingly, the human meridian system may serve as an illustrative example of how structural networks embody the recurring organizational principles described by the six structural laws.

VII. A Modern Structural Interpretation

This book proposes that **Point •, Line ｜, and Circle O** should not be regarded as repetitions of ancient philosophical ideas.

Instead, they constitute a modern structural language in which:

- **Point •** represents localized generation;
- **Line ｜** represents connection and evolution;
- **Circle O** represents structural closure.

These symbols are not natural objects. They represent the three universal structural roles shared by all stable systems.

Accordingly, Modern Structural Language provides a unified framework through which the common structural features found in diverse traditional knowledge systems may be reinterpreted.

Traditional thought relied primarily upon:

- observation;
- experience;
- holistic intuition.

Modern science relies upon:

- experimentation;
- mathematics;
- testable models.

Although their methodologies differ, both seek to answer the same fundamental question: **Why is the complex world capable of maintaining order?**

Therefore, this book does not reject traditional wisdom. Rather, it attempts to express its universally meaningful structural insights using the language of modern science.

VIII. Summary

Different civilizations have developed different symbolic languages. Yet all have attempted to identify the universal structural principles underlying reality.

Point •, Line ｜, and Circle O provide a modern structural language capable of integrating:

- traditional holistic thinking,
- modern scientific theory,
- complexity science,

within a single conceptual framework.

Accordingly, they belong not exclusively to mathematics, physics, or chemistry, but function as a common bridge connecting **Ancient Structural Thinking** with **Modern Scientific Language**.

Structural Conclusion:

Ancient civilizations sought universal principles through symbolic structural thinking. Modern science seeks universal principles through mathematical models.

The framework of **Point •, Line ｜, and Circle O** does not replace traditional wisdom. Rather, it provides a common structural language through which ancient holistic thought and modern scientific theory may be understood within the same conceptual framework.

Methodological Significance:

The theoretical advancement presented here is even more significant than earlier interpretations. Previously, one might simply have suggested that the **Five Elements resemble Point •, Line ｜, and Circle O**.

This book proposes something fundamentally different.

The **Five Elements**, **Yin–Yang**, the **Three Powers**, and **Traditional Chinese Medicine** belong to **Ancient Structural Thinking**. **Point •, Line ｜, and Circle O** belong to **Modern Structural Language**.

Neither framework serves to prove or derive the other. Rather, both represent distinct intellectual traditions that seek to understand the complexity of the world through different modes of expression.

Accordingly, the **Structural Periodic Table** should be understood not as a reinterpretation of ancient philosophy, but as the proposal of a new scientific language capable of communicating with both traditional wisdom and modern science within a unified structural framework.

Section 6. From Natural Cycles, Newtonian Mechanics, and Einsteinian Geometry to Structural Periodicity

Abstract:

A unifying pattern gradually emerges throughout *Structural Periodic Table*: **the lower the structural level, the more evident periodicity becomes; the higher the structural level, the more dominant structural organization becomes.**

At the chemical level, periodicity is expressed through the periodic recurrence of elemental properties. At higher levels of biological organization, strict periodicity gradually gives way to recurring structural patterns. In astronomy and cosmology, scientific attention has likewise shifted from periodic motions to geometric organization and large-scale structural formation.

From this perspective, the history of natural science itself reflects a common evolutionary trajectory: **Periodicity → Geometry → Structure**

Correspondingly, scientific inquiry has progressed from describing **motion**, to explaining **spacetime geometry**, and ultimately toward understanding **the formation of stable structures**.

The *Structural Periodic Table* does not attempt to replace classical mechanics or relativity. Rather, it proposes a complementary structural framework for investigating why stable hierarchical structures repeatedly emerge across different scales of nature.

I. The Chemical Level

The clearest example of periodicity is the **Periodic Table of Elements**.

H → He; Li → Ne; Na → Ar ...

The recurrence of chemical properties is remarkably regular, allowing even undiscovered elements to be predicted. At this level, **periodicity dominates.**

II. The Biological Level

Life displays a different pattern: **Virus → Bacterium → Multicellular Life → Animals**

Here, strict periodicity gradually weakens. Instead, recurring **structural organization** becomes increasingly apparent.

Typical examples include: **Cell membranes; DNA; Neural networks**

These structures repeatedly appear across increasingly complex biological systems.

III. The Plant Level

Plant development also illustrates this transition: **Seed → Germination → Growth → Flowering → Fruiting → Seed**

Although this appears cyclic, modern botany focuses less on seasonal repetition than on structural organization, including:

- Branching architecture
- Leaf venation
- Fractal geometry
- Golden-angle phyllotaxis
- Root network organization

Thus, plant science likewise evolves from the study of **cycles** toward the study of **structures**.

IV. The Astronomical Level

The same tendency becomes even more evident in astronomy.
Early astronomy primarily investigated:

- Day and night
- Lunar phases
- Seasons

These are all periodic phenomena.

Newton later described orbital motion through

$$F = \{GMm\}/\{r^2\}$$

placing emphasis on orbital dynamics.

Einstein subsequently introduced

$$G_{\{\mu\nu\}} = \frac{\{8\pi G\}}{\{c^4\}} T_{\{\mu\nu\}}$$

shifting scientific attention from periodic motion to the geometry of spacetime itself.

Modern cosmology now concentrates on structures such as:

- The cosmic web
- Galaxy clusters
- Dark matter distribution
- Gravitational lensing

The scientific focus has therefore evolved from **motion** toward **structure**.

Discipline	**Earlier Focus**	**Current Focus**
Chemistry	Periodicity	Structural organization
Life Science	Life cycles	Biological structure
Botany	Growth cycles	Fractal architecture
Zoology	Reproductive cycles	Neural organization
Astronomy	Orbital cycles	Cosmic structure

Table 3-12-6-1: Scientific Attention: From Cycles to Structures

The *Structural Periodic Table* therefore does not simply imitate the Periodic Table of Elements. Instead, it proposes that **periodicity**

is the observable manifestation of structure, whereas structure is the underlying cause of periodicity.

Across natural science, the general developmental trend appears to be **Periodic Phenomena → Geometric Organization → Structural Formation**

or, equivalently, **Equations of Motion → Geometric Equations → Structural Equations**

Future studies of plants, animals, astronomy, and other complex systems may continue along this same trajectory. Within the Structural Periodic Table, this corresponds to the continual recurrence of **Point • → Line | → Circle O**

across different disciplines and hierarchical scales.

V. Toward Structural Periodicity

Historically, enduring scientific theories share one important characteristic: they successfully explain previously unresolved phenomena.

Theory	**Principal Achievement**
Newtonian Mechanics	Planetary motion
Maxwell's Electromagnetism	Electromagnetic waves
General Relativity	Mercury's perihelion advance; gravitational lensing
Quantum Mechanics	Atomic spectra
Periodic Table	Prediction of undiscovered elements

Table 3-12-6-2: Enduring Scientific Theories

This naturally raises an important question:

If the Structural Periodic Table is more than an interpretive framework, what kinds of phenomena should it be able to predict?

5.1 Newton and Einstein

Many problems are already successfully explained. For example,

- planetary motion around the Milky Way,
- Mercury's perihelion advance,
- black-hole mergers.

These lie within the established domains of classical mechanics and general relativity.

The Structural Periodic Table does not seek to replace these theories.

5.2 Structural Formation

Its potential strength lies elsewhere. Rather than asking **How does a structure move?**

it asks **Why does this structure emerge in the first place?**

Newton explains gravitational motion, but not why the Solar System possesses its particular hierarchical organization.

Einstein explains spacetime curvature, but not why spiral galaxies repeatedly develop stable spiral-arm configurations.

These questions belong naturally to structural formation.

5.3 Hierarchical Organization of the Solar System

Examples include:

- Why is the number of planets limited?
- Why do orbital zones appear discretized?
- Why do asteroid belts occupy particular regions?

These questions resemble those asked in chemistry: Why are chemical families finite?

One may therefore consider the possibility of **hierarchical orbital periodicity** within planetary systems.

5.4 Spiral Galaxies

Many spiral galaxies possess

- two spiral arms,
- three spiral arms,
- four spiral arms.

Why? Why do some approach the golden angle while others deviate?

Traditional explanations rely primarily on

- density-wave theory,
- numerical simulations.

Structural Periodicity instead investigates **which structural configurations are stable and which are unstable.**

5.5 Large-Scale Cosmic Structure

Modern cosmology studies

- galaxies,
- galaxy clusters,
- superclusters,
- the cosmic web.

Standard cosmology explains these as consequences of primordial fluctuations. Structural Periodicity asks additional questions:

- Why do hierarchical levels emerge?
- Why are they finite?
- Why do stable structural windows exist?

These questions naturally correspond to the concepts of **closure, stability, and upper limits** developed throughout this book.

Accordingly,

$$\{dM\}/\{dt\} = \alpha_1 \nabla M + \alpha_2 I(E, S, C) + \alpha_3 Q(x)$$

is proposed as a structural equation describing **structural formation**, rather than orbital calculation.

Newton investigated **position.** Einstein investigated **spacetime.** The Structural Periodic Table attempts to investigate **structure itself.**

Its central question is therefore: **Why does the universe repeatedly organize itself into stable hierarchical structures?**

This establishes a coherent intellectual progression: **Newtonian Motion → Einsteinian Geometry → Structural Formation**

Era	Scientific Focus	Representative Theory
Natural Observation	Cycles	Ancient Astronomy, Calendar Systems
Classical Physics	Motion	Newtonian Mechanics
Modern Physics	Geometry	General Relativity
Structural Science	Formation	Structural Periodicity

Table 3-12-6-3: Evolution of Scientific Paradigms

Scientific Era	Fundamental Question
Ancient Science	What repeats?
Newton	How does it move?
Einstein	How does spacetime curve?
Structural Periodicity	Why do stable structures form?

Table 3-12-6-4: Central Questions of Scientific Development

Concluding Statement:

Newton explained how bodies move. Einstein explained how spacetime bends. Structural Periodicity attempts to explain why stable structures repeatedly emerge across different scales of nature.

Motion describes change. Geometry describes relationships. Structure describes organization. Together, they represent successive stages in the evolution of scientific thought.

References (Chapter 12)

Philosophy of Science

[1] Thomas S. Kuhn (2012). *The Structure of Scientific Revolutions* (50th Anniversary Edition). University of Chicago Press.

[2] Karl Popper (2002). The Logic of Scientific Discovery. Routledge.

[3] Karl Popper (1972). Objective Knowledge. Oxford University Press.

Mathematics and Structure

[4] Hermann Weyl (1952). Symmetry. Princeton University Press.

[5] Timothy Gowers (Ed.). (2008). The Princeton Companion to Mathematics. Princeton University Press.

Physics

[6] Richard P. Feynman, Leighton, R. B., & Sands, M. (2011 New Millennium Edition). The Feynman Lectures on Physics. Basic Books.

[7] Albert Einstein (2001). Relativity: The Special and the General Theory. Routledge.

Complexity and Systems

[8] Melanie Mitchell (2009). Complexity: A Guided Tour. Oxford University Press.

[9] Albert-László Barabási (2016). Network Science. Cambridge University Press.

[10] Herbert A. Simon (1962). The Architecture of Complexity. *Proceedings of the American Philosophical Society*, 106(6), 467–482.

[11] Steven Strogatz (2003). Sync. Hyperion.

Structural Thinking

[12] D'Arcy Wentworth Thompson (1992 reprint). On Growth and Form. Cambridge University Press.

[13] Stuart A. Kauffman (1993). The Origins of Order. Oxford University Press.

[14] Erwin Schrödinger (1944). What Is Life?. Cambridge University Press.

Classical Chinese Structural Thought

[15] Tao Te Ching.

[16] I Ching.

[17] Huangdi Neijing.

Structural Periodicity Theory

[18] Chang, J. (2013). *Grand Ultimate Theory*. Universal Publishing.

[19] Chang, J. (2026). *Recursive Self-Organizing Evolutionary Breakthrough: Unifying Natural, Social and Life Sciences*. Universal Publishing.

[20] Chang, J. (2026). *Metastructural Unification (Physics Volume): Unifying Electromagnetism, Gravitation and Quantum Theory*. Universal Publishing.

[21] Chang, J. (2026). *Ultra-Metastructural Unification (Mathematics Volume): Unifying Algebra, Geometry and Probability*. Universal Publishing.

[22] Chang, J. (2026). *Metastructural Unification (Chemistry Volume): Unifying Reactions, Compounds and Elements*. Universal Publishing.

Chapter Thirteen: The Predictive Framework of the Structural Periodic Table — From Known Structures to Unknown Structures

Figure 3-13: Crop circle illustrating Galactic Civilization. See book 《*Crop Circle*》 for details.

Chapter Overview:

A scientific theory should do more than explain phenomena that have already been observed; it should also possess the capacity to predict structures that have not yet been discovered.

The Periodic Table of Elements became one of the greatest achievements in modern science not merely because it organized known elements, but because it successfully predicted unknown ones that were later confirmed experimentally. If the **Structural Periodic Table** is to become a genuine scientific framework, it must possess a comparable predictive capability.

The predictions discussed in this book, however, differ fundamentally from conventional scientific prediction. They are not intended to predict a particular numerical value or the outcome of a specific experiment. Instead, they seek to predict:

- Which structural configurations are capable of achieving long-term stability;

- Which theoretical systems still contain structural vacancies;
- Which emerging disciplines possess structural inevitability;
- Which structures cannot persist because they violate the conditions of structural closure.

Consequently, this chapter is concerned less with **predicting the future** than with **developing a methodology for structural prediction**.

It first establishes the fundamental principles of structural prediction, then discusses methods for identifying structural vacancies, and finally proposes several testable examples of structural forecasting, thereby preparing the discussion of theoretical boundaries and structural adjudication presented in the next chapter.

Section 1. Why Does the Structural Periodic Table Possess Predictive Power?

Abstract:

One of the central questions addressed in this book is whether the **Structural Periodic Table** possesses genuine predictive capability and whether its framework can be extended beyond currently established knowledge.

To address this question, we begin by constructing the theoretical foundation of the Structural Periodic Table from the perspective of the fundamental sciences and then examine the mechanisms underlying structural prediction.

I. Core Structural Framework

The Periodic Table of Elements reveals the recurrence of **periodicity**.

The Structural Periodic Table proposes the recurrence of **structure**.

As system complexity increases, strict periodicity gradually diminishes, while structural repetition becomes increasingly dominant.

1.1 The Structural Triad

Point • — Source of generation
Line ｜ — Evolutionary pathway
Circle O — Structural closure

1.2 From Elements to Life

Throughout this book, the following structural sequence has been established: **Elements → Reactions → Configurations → Life**

corresponding to **Point • → Line ｜ → Circle O → Living Systems**

This sequence represents the progressive emergence of increasingly complex structural organization.

II. The Significance of the Structural Periodic Table

The Periodic Table explains the material world while predicting undiscovered chemical elements.

The Structural Periodic Table seeks to explain — and ultimately predict — unknown structures, including:

- The emergence of scientific disciplines;
- The formation of theoretical systems;
- The origin of life;
- The evolution of complex systems.

Its objective is therefore not merely classification, but the discovery of general principles governing structural formation across multiple levels of organization.

III. Beyond a Descriptive Framework

Earlier chapters began with the Periodic Table of Elements and developed the theoretical foundations of the Structural Periodic Table. Mathematics, physics, and chemistry were subsequently examined as representative examples demonstrating how structural periodicity manifests itself throughout the natural sciences.

The discussion then extended to the emergence of life from chemical organization, thereby establishing a continuous structural progression from matter to living systems.

However, a theory that aspires to become a genuine scientific framework must ultimately address two additional questions:

- **What can it predict?**
- **What are its limitations?**

The development of a mature scientific theory generally proceeds through three successive stages.

Stage One: Structural Construction

- Discover structural regularities;
- Establish a unified structural language;
- Construct a coherent theoretical framework.

Stage Two: Structural Extrapolation

- Predict unknown structures from known structural principles;
- Identify structural vacancies within existing frameworks;
- Infer future directions of scientific development through structural analysis.

Stage Three: Structural Delimitation

- Define the domain of applicability;
- Determine the conditions under which the theory fails;
- Establish objective criteria for theoretical adjudication.

The present chapter occupies the transition between the second and third stages.

IV. The predictive capability

Its primary concern is therefore not the introduction of additional scientific content, but rather the question: **Does the Structural Periodic Table possess genuine predictive capability?**

To answer this question, the following unified structural criteria are proposed:

- **Periodicity** — the recurrence of structures after structural closure;
- **Hierarchy** — the increase of structural degrees of freedom across successive organizational levels;
- **Adjudication** — the emergence and selection of stable structural windows.

Only under a unified set of structural criteria can predictions become both comparable and scientifically testable.

Throughout this book, three foundational scientific disciplines serve as the principal structural axis:

- **Mathematics** — the language of structure;
- **Physics** — the dynamics of nature;
- **Chemistry** — the organization of matter.

Together they constitute the fundamental triangular framework upon which the Structural Periodic Table is built.

Higher-level systems — including the life sciences and the social sciences — may subsequently be regarded as further hierarchical expansions of this foundational structure.

Consequently, the central question explored throughout this chapter is: **If the Structural Periodic Table correctly describes the mechanisms of structural formation, can it also predict future theories, disciplines, and structural organizations in the same way that the Periodic Table successfully predicted previously unknown elements?**

The following sections examine this question from multiple perspectives.

Section 2. From Elemental Vacancies to Structural Vacancies

Abstract:

This book proposes that if the emergence of scientific disciplines follows a unified mechanism of structural generation, then the **Structural Periodic Table** is not merely a classification system for existing disciplines but also a predictive framework for future scientific development.

By analyzing the structural closure of the **Point–Line–Circle** triad across multiple hierarchical levels, we propose that every apparent "vacancy" within the Structural Periodic Table represents either an underdeveloped or as-yet unnamed structural domain.

This section establishes the principles, criteria, and practical methodology of structural prediction while clarifying how structural prediction fundamentally differs from conventional empirical forecasting.

I. From the Periodic Table of Elements to the Structural Periodic Table

The greatness of the Periodic Table lies not merely in classification but in prediction.

Dmitri Mendeleev did not simply organize known elements. Instead, he:

- discovered structural periodicity;
- recognized structural vacancies;
- predicted the necessary existence of unknown elements.

The later discoveries of gallium, germanium, and scandium demonstrated that **structural order preceded empirical discovery.**

The Structural Periodic Table attempts to accomplish a similar objective, except that its subject changes from **chemical elements** to **scientific structures**.

Periodic Table of Elements	Structural Periodic Table
Nuclear charge determines electron shells	Structural equations determine disciplinary organization
Vacancies indicate undiscovered elements	Vacancies indicate undiscovered structural domains
Predicts matter	Predicts theoretical directions

Table 3-13-2-1: Predictive Comparison Between the Periodic Table and the Structural Periodic Table

---

II. The Structural Generative Mechanism

The central equation proposed throughout this book,

$$dM/dt = \alpha_1 \cdot \nabla M + \alpha_2 \cdot I(E,S,C) + \alpha_3 \cdot Q(x)$$

is not intended as a discipline-specific equation. Instead, it represents a general mechanism of structural generation. Within this framework,

- **∇M** represents the emergence of local structural differentiation;
- **I(E,S,C)** represents structural integration and stable closure;
- **Q(x)** represents directional structural generation.

A scientific discipline can achieve long-term stability only when these three structural processes form a complete closure.

III. Identifying Undiscovered Structural Domains

Structural prediction is not speculation. It is a process of structural diagnosis.

3.1 Criterion I: The Principle of Structural Vacancies

Whenever a particular hierarchical level contains **Point, Line, and Circle** structures but lacks the corresponding fourth-level feedback or integrative structure, an additional interdisciplinary framework is likely to remain undeveloped.

For example, mathematics has already established algebra, geometry, and probability.

Yet a fully unified theory of structural generation remains incomplete. Such incompleteness represents a **structural vacancy**.

3.2 Criterion II: The Principle of Cross-Disciplinary Asymmetry

When different disciplines reach different levels of structural development, asymmetry naturally appears. For example,

- mathematics may have completed its fourth structural level;
- physics may have completed its fourth structural level;
- chemistry may still remain primarily within the third level.

Such asymmetry suggests that a fourth-level structural chemistry represents a natural future direction.

3.3 Criterion III: The Principle of Structural Imbalance

A scientific field frequently exhibits structural incompleteness whenever it possesses:

- abundant technology,
- fragmented explanations,
- competing theoretical models,
- no unified conceptual language.

Such regions are precisely where future disciplines are most likely to emerge.

IV. What Types of New Structures Can Be Predicted?

4.1 Cross-Hierarchical Structural Disciplines

Possible examples include:

- Structural Information Physics
- Structural Statistical Cosmology
- Structural Generative Chemistry
- Structural Biodynamics

4.2 Structural Adjudication Sciences

These disciplines would not study particular objects. Instead, they would investigate **which theoretical structures are structurally admissible.** Examples include:

- Structural Theory of Existence
- Structural Consistency Theory

4.3 Artificial Intelligence and Structural Science

Because the Structural Periodic Table is inherently algorithmic, future developments may include:

- automated structural reasoning systems;
- structural closure verification algorithms;
- automatic identification of structural vacancies.

These possibilities illustrate the natural compatibility between structural theory and artificial intelligence.

V. Structural Prediction versus Empirical Prediction

Traditional scientific forecasting generally relies upon:

- technological trends;
- experimental progress;
- social demands.

Structural prediction instead relies upon:

- structural completeness;
- hierarchical symmetry;
- structural balance.

The former represents **statistical forecasting**.

The latter represents **structural necessity.**

A particularly important implication follows: If the Structural Periodic Table represents a complete structural framework, then future scientific disciplines are not merely possible — they are simply awaiting discovery and formal recognition.

VI. The Ultimate Comparison with the Periodic Table

Atomic Theory	Structural Theory
Nuclear charge determines elements	Structural equations determine disciplines
Electron shells determine periodicity	Structural rings determine hierarchy
Vacancies → New elements	Vacancies → New theories
Experimental verification	Theoretical and AI-assisted verification

Table 3-13-2-2: Atomic Theory and Structural Theory

VII. Conclusion

The Periodic Table predicts **matter**. The Structural Periodic Table attempts to predict **structures of knowledge.**

Its objective is therefore not simply to extend existing science, but to explain **how scientific structures themselves emerge.**

From this perspective, scientific discovery becomes a structural process rather than merely an accumulation of empirical observations.

Structural Conclusion:

The greatest achievement of the Periodic Table was not the organization of known elements, but the prediction of unknown ones. Likewise, the ultimate value of the Structural Periodic Table lies not in reorganizing existing disciplines, but in revealing structural vacancies that point toward future theories, emerging disciplines, and previously unrecognized forms of organization. Prediction, therefore, is not an act of speculation, but a consequence of structural necessity.

Section 3: Three Possible Theoretical Breakthroughs of the Future

Abstract:

If scientific structures follow the principle of **Point • – Line | – Circle O × Three-Ring–Four-Ring Structural Closure**, then the current scientific system should contain a number of **structural vacancies**. Based on the closure criteria proposed by the **Structural Periodic Table**, this section identifies three theoretical directions that are most likely to emerge in the future:

- **Theory of Structural Emergence** *(Line Structure)* — e.g., **Structural Information Emergence**;
- **Theory of Structural Adjudication** *(Point Structure)* — e.g., **Structural Adjudication Physics**;
- **Cross-Level Stability Theory** *(Circle Structure)* — e.g., **Cross-Level Stability Science**.

These potential breakthroughs are not extrapolated from technological trends. Instead, they arise from the intrinsic necessity to restore **structural imbalance** and **complete missing closure** within the scientific knowledge system.

I. Prediction Is Not Prophecy, but Structural Completion

The most successful scientific predictions in history have not been acts of speculation, but consequences of **structural incompleteness**. Examples include:

- **Mendeleev** predicted gallium and germanium from vacancies in the Periodic Table.
- **Dirac** predicted antimatter because his equation required it.
- **Higgs** predicted the Higgs boson to complete the symmetry structure of particle physics.

In each case, a **structural vacancy** generated theoretical tension. When that tension became sufficiently strong, a scientific breakthrough followed.

II. Three Structural Imbalances in Contemporary Science

A comparison of the structural organization of **mathematics, physics, chemistry, and information science** reveals three major structural deficiencies.

2.1 Missing Theory of Structural Emergence

Today we are able to:

- describe structures;
- simulate complex systems;
- perform large-scale computation and statistical analysis.

Yet we still cannot answer a more fundamental question: **Under what structural conditions must complex organization inevitably emerge?**
For example:

- Artificial intelligence can generate novel content;
- The universe can organize galaxies;
- Biological systems can self-organize.

However, there is still no unified **criterion for structural emergence**.

2.2 Missing Theory of Structural Adjudication

Modern theoretical physics contains numerous competing models:

- modified gravity theories;
- alternative cosmological models;
- multiple interpretations of quantum mechanics.

Yet there is no **structural adjudication mechanism** capable of determining which theoretical structures are fundamentally admissible. Current evaluation standards rely mainly on:

- computational consistency;
- observational agreement;
- mathematical elegance or symmetry.

What remains absent is a criterion based on **structural closure itself**.

2.3 Missing Cross-Level Stability Theory

Scientific knowledge remains fragmented across different organizational levels:

- quantum ↔ macroscopic physics;
- chemistry ↔ life;
- local systems ↔ the universe.

No unified discipline currently investigates **how structural stability is maintained across different hierarchical levels**.

This missing framework may ultimately become **Cross-Level Stability Theory**.

III. Returning Prediction to Structural Periodicity

Predictions within the **Structural Periodic Table** are not based on technological trends or historical extrapolation. They are derived from three structural criteria.

Criterion 1: Hierarchical Vacancy

If a particular level has not yet completed **four-ring structural closure**, that missing level will eventually be completed.

This structural vacancy naturally gives rise to the **Theory of Structural Emergence**, representing the **Line Structure**.

Criterion 2: Structural Imbalance

When numerous competing models exist without a unified structural criterion for selecting among them, a new discipline devoted to structural adjudication becomes inevitable.

This leads naturally to the **Theory of Structural Adjudication**, representing the **Point Structure**.

Criterion 3: Cross-Disciplinary Asymmetry

When one scientific discipline advances substantially beyond others in structural completeness, a higher-level interface is required to restore balance among disciplines.

This structural necessity leads to the **Cross-Level Stability Theory**, representing the **Circle Structure**.

These three structural principles together suggest **three possible theoretical breakthroughs** that may shape the future development of science.

IV. Theory of Structural Emergence

4.1 Predicted Theory I

1) Predicted Development

A new theoretical framework may emerge to answer the following questions:

- Under what structural constraints do complex patterns necessarily emerge?
- Which systems are capable of self-organization?
- Which systems cannot achieve self-organization?

Such a theory could provide a unified explanation for:

- spectral emergence in random matrix theory;
- self-organization in living systems;

- structural emergence in artificial intelligence.

2) Why Is It Inevitable?

At present,

- we possess numerous generative models,
- but we lack a theory capable of determining **when structural generation is necessary rather than merely possible**.

In other words, the technology has matured, whereas the underlying theoretical framework has not yet achieved structural closure.

4.2 Predicted Example I: Structural Information Emergence

Definition: *Structural Information Emergence* is proposed as a future discipline devoted to understanding **how information is necessarily generated under conditions of structural closure.**

1) Why Should Such a Discipline Emerge?

Currently,

- Information Theory studies coding and entropy;
- Complexity Science investigates emergence;
- Artificial Intelligence develops generative models;
- Physics studies spectral structures.

However, no existing discipline asks the deeper structural question: **Under what structural conditions must information emerge, rather than simply being capable of emerging?**

From the perspective of the Structural Periodic Table, this represents a genuine structural vacancy.

- Mathematics possesses theories of generation.
- Physics studies spectral emergence.
- Chemistry investigates structural formation.

• Yet Information Science still lacks a **criterion for structural emergence.**

2) Research Topics

Possible research topics include:

• how high-dimensional spectra transform into information structures;
• why certain complex systems necessarily generate stable patterns;
• why AI models exhibit self-organization only at particular structural levels.

3) Potential Applications

Possible applications include:

• controllable structural generation in AI;
• pattern formation in neuroscience;
• information-based interpretations of cosmic structural evolution.

4.3 Additional Prediction Case I: A Structural Interpretation of Spectral Repulsion

As an initial example of structural prediction, this book proposes that spectral repulsion in complex quantum systems is not merely a statistical consequence of random matrices, but instead arises from three fundamental structural conditions:

• **Point • — Irreducibility**
• **Line | — Interaction Connectivity**
• **Circle O — Global Structural Closure**

This interpretation suggests that the **Wigner–Dyson statistics** of Random Matrix Theory may be understood as a statistical manifestation of completed structural organization rather than as a fundamental assumption of randomness.

Should future investigations confirm this interpretation across broader classes of high-dimensional many-body systems, it would represent an important demonstration of the predictive capability of the Structural Periodic Table.

More importantly, it would illustrate that structural principles can not only explain existing theories but also generate entirely new theoretical frameworks.

This case also serves as one of the first practical demonstrations of **structural prediction**, providing a natural foundation for incorporating additional prediction cases in future editions.

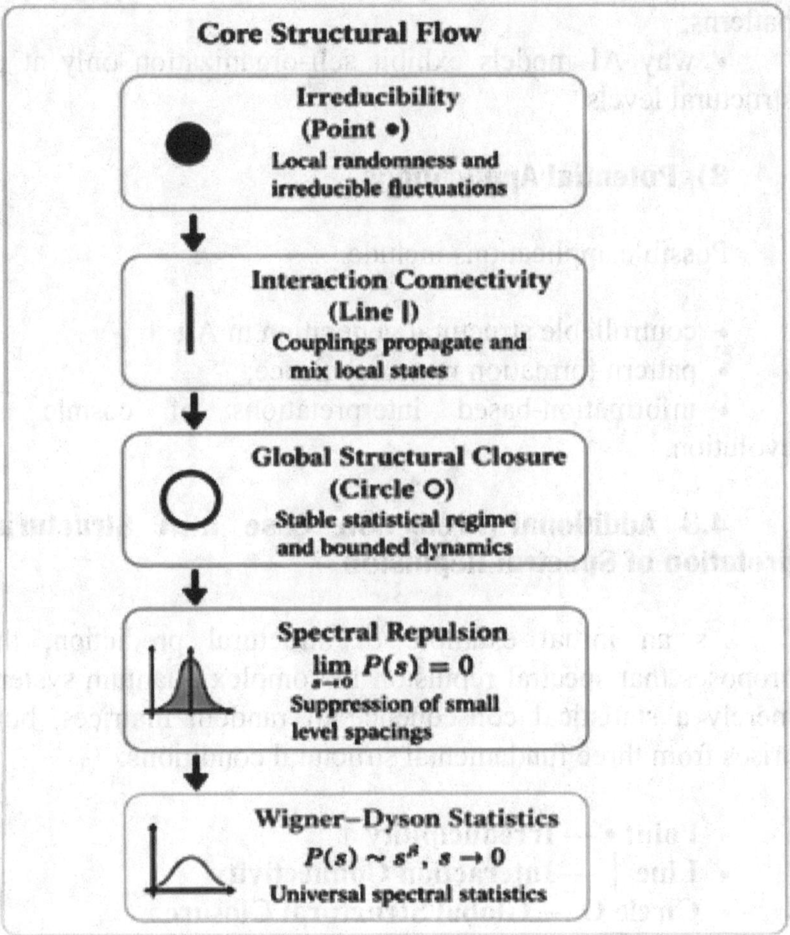

(1)

8. Numerical Illustrations (Toy Simulations)

(a) 2×2 Hermitian Matrix (Analytical)

Consider $H = \begin{pmatrix} a & b \\ b^* & c \end{pmatrix}$ (with $b \in \mathbb{C}$).

Eigenvalues: $E_\pm = \frac{a+c}{2} \pm \sqrt{\left(\frac{a-c}{2}\right)^2 + |b|^2}$

Level spacing: $s = E_+ - E_- = 2\sqrt{\left(\frac{a-c}{2}\right)^2 + |b|^2}$

Parameters	Level spacing s
$a = c = 0, \quad b = 0$	$s = 0$ (degenerate)
$a = c = 0, \quad b = 0.1$	$s = 0.2$
$a = 0.2, c = 0, \ b = 0.1$	$s \approx 0.2236$

(b) 10×10 GOE Matrix (Numerical)
Ensemble: Gaussian Orthogonal Ensemble (GOE), $N = 10$.
We generate 10,000 random matrices and compute the nearest-neighbor spacing distribution.

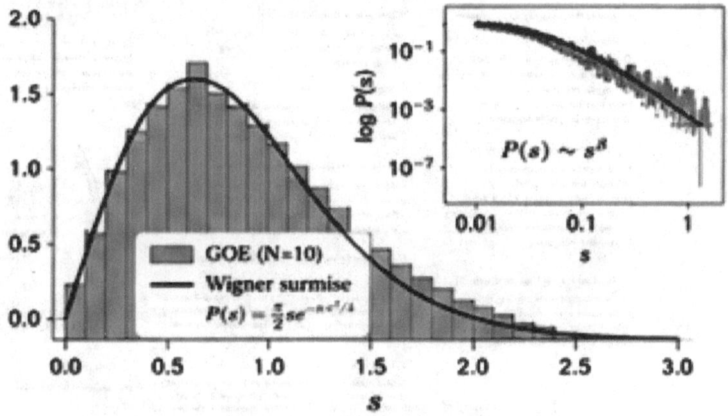

Observation: $P(s)$ **vanishes linearly as** $s \to 0$ **($\beta \approx 1$ for GOE), confirming spectral repulsion.**

(2)

(3)

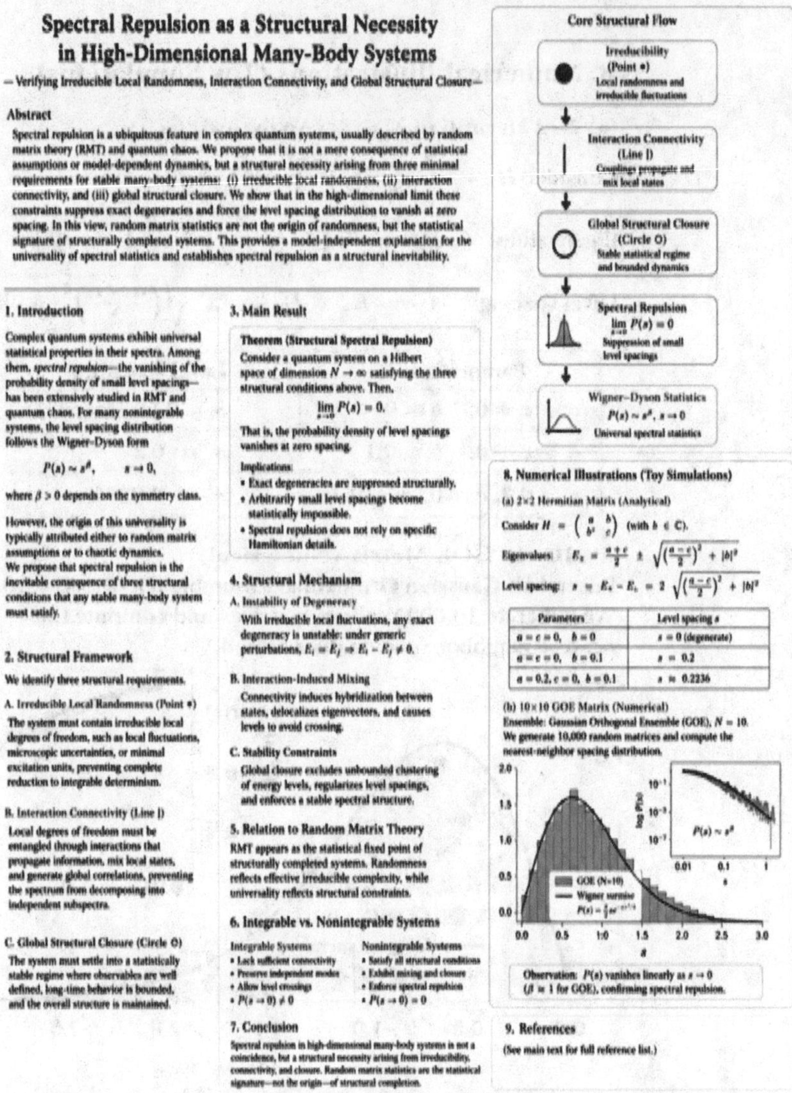

Figure 3-13-3-1. Spectral Repulsion as a Structural Necessity in High-Dimensional Many-Body Systems

(Verifying local irreducibility, interaction connectivity, and global structural closure.)

V. Theory of Structural Adjudication

5.1 Predicted Theory II

1) Predicted Development

A new form of theoretical physics may emerge whose primary purpose is **not to solve equations, but to determine which theoretical structures are structurally admissible.**

Such a framework could influence areas including:

- Multiverse theories;
- String Theory;
- Modified gravity;
- Dark energy models.

2) Why Is It Inevitable?

The number of theoretical models continues to grow rapidly, yet there is still no structural mechanism capable of selecting among them.

From the perspective of the Structural Periodic Table, the **Fourth-Ring adjudication mechanism** has not yet been systematically established.

5.2 Predicted Example II: Structural Adjudication Physics

Definition: Structural Adjudication Physics is proposed as a discipline that investigates **which theoretical equations and structures are permitted to exist**, rather than how particular equations should be solved.

1) Why Should Such a Discipline Emerge?

Contemporary physics contains

- numerous competing theories (String Theory, the Multiverse, modified gravity, etc.);
- increasingly sophisticated mathematical techniques;

- highly fragmented interpretations.

Yet it lacks a unified structural language for theoretical selection. Just as the Periodic Table determines which elements are permitted within chemical structure, physics likewise requires principles capable of answering:

- Which theoretical structures can achieve closure?
- Which violate structural generation?
- Which destroy structural stability?

These questions define the scope of Structural Adjudication Physics.

2) Research Topics

Its research may include:

- criteria for theoretical admissibility;
- structural selection among competing theories;
- whether higher-dimensional models satisfy structural closure.

3) What It Does Not Attempt

Structural Adjudication Physics does **not** aim to:

- derive specific partial differential equations;
- compute detailed spectra;
- construct new particle models.

Instead, its objective is to determine **whether a theoretical structure is structurally permissible.**

5.3 Additional Prediction Case II

Possible future applications include structural adjudication analyses of

- String Theory,

- M-Theory,
- Loop Quantum Gravity,

all of which have been discussed in earlier chapters.

VI. Cross-Level Stability Theory

6.1 Predicted Theory III

1) Predicted Development

A unified theory may eventually emerge to explain:

- why macroscopic entropy increase does not destroy microscopic reversibility;
- why living systems maintain long-term stability;
- why large-scale cosmic structures remain stable over cosmological time.

Such a theory could unify

- statistical physics;
- complexity science;
- cosmology;
- biological organization.

2) Why Is It Inevitable?

Modern science is increasingly integrating multiple scales:

- quantum information;
- quantum biology;
- cosmic evolution.

However, no unified language yet exists to explain **cross-level structural stability.**

6.2 Predicted Example III: Cross-Level Stability Science

Definition: Cross-Level Stability Science is proposed as a discipline devoted to understanding how structural consistency is maintained across multiple hierarchical levels.

1) Why Should Such a Discipline Emerge?

Reality contains numerous transitions:

- quantum → macroscopic;
- molecules → life;
- local systems → the universe.

Yet no unified discipline currently investigates how stable organization persists across these transitions.

Within our framework, this problem is naturally represented by the structural equation

$$dM/dt = \alpha_1 \cdot \nabla M + \alpha_2 \cdot I(E,S,C) + \alpha_3 \cdot Q(x)$$

which describes how structural closure is maintained across different hierarchical levels.

2) Research Topics

Possible research topics include:

- reconciling macroscopic entropy growth with microscopic reversibility;
- understanding the transition from chemical structure to living systems;
- maintaining stability within social systems.

3) Significance

Cross-Level Stability Science may ultimately become a unifying bridge connecting

- physics,

- chemistry,
- life sciences,
- and the social sciences.

6.3 Additional Prediction Case III

An interesting example concerns **three-dimensional time**.

In Chapter 9 of the author's *Universal Law* (published between **2003 and 2006**), the concept of **three-dimensional time corresponding to three-dimensional space** was proposed and interpreted primarily from a macroscopic perspective.

More recently, **Gunther Kletetschka** proposed a mathematical formulation of three-dimensional time in the paper *Three-Dimensional Time: A Mathematical Framework for Fundamental Physics* (2025), approaching the concept from the microscopic and mathematical perspective.

Although the two works differ substantially in methodology, motivation, and theoretical framework, they illustrate how similar structural ideas may emerge independently within different scientific contexts. This example suggests that structural concepts proposed earlier may later receive new mathematical formulations through independent developments.

VII. Structural Origins of the Three Predicted Disciplines

Structural Vacancy	Predicted Discipline
Missing emergence mechanism	Structural Information Emergence
Missing theoretical selection mechanism	Structural Adjudication Physics
Missing cross-level consistency	Cross-Level Stability Science

Table 3-13-3-1: Structural Vacancies and Predicted Future Disciplines

When structural closure remains incomplete, scientific breakthroughs are not accidental — they represent the natural completion of the structural system. New disciplines therefore arise not merely by choice, but by structural necessity.

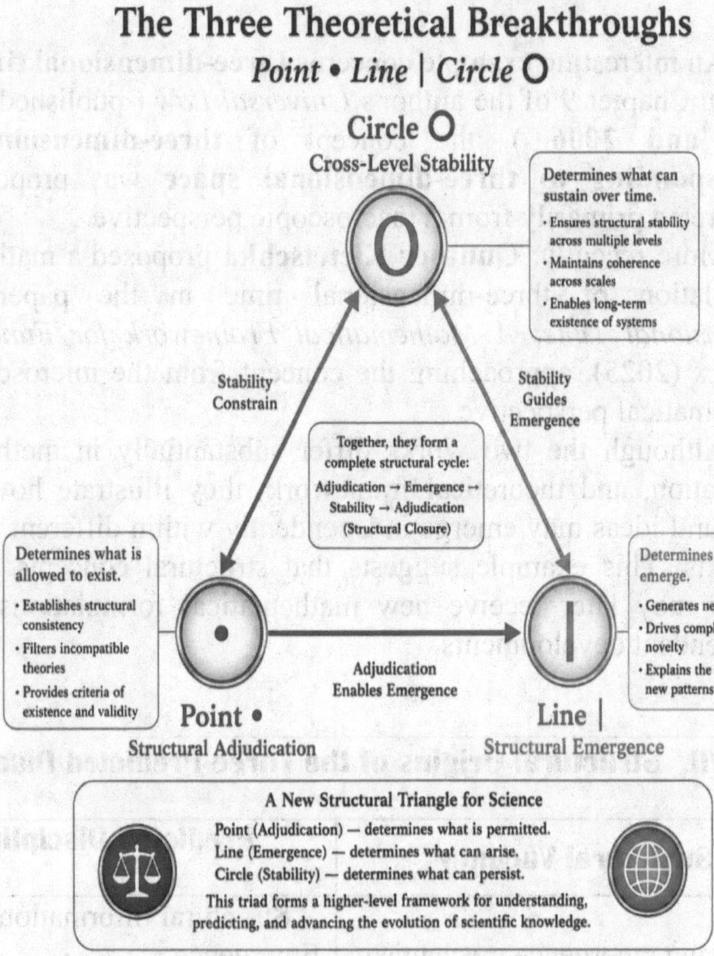

Figure 3-13-3-2: Three Classes of Theoretical Breakthroughs ----- Point •, Line ｜, and Circle O

VIII. Ultimate Comparison with the Periodic Table

Periodic Table	Structural Periodic Table
Predicts unknown elements	Predicts unknown theories
Structural vacancies produce new elements	Structural vacancies produce new disciplines
Verified experimentally	Verified theoretically and assisted by AI

Table 3-13-3-2: The Periodic Table and the Structural Periodic Table

The Periodic Table predicts the future of the **material world.**
The Structural Periodic Table seeks to predict the future of the **knowledge world.**

Concluding Statement:

The Periodic Table predicts vacancies in matter; the Structural Periodic Table predicts vacancies in knowledge.
In the foreseeable future, possible breakthroughs may include:

- a mathematical framework for **Structural Emergence Theory**;
- the incorporation of **Structural Adjudication Theory** into mainstream theoretical physics;
- the establishment of **Cross-Level Stability Science** as a central interdisciplinary field.

Therefore, the purpose of the **Structural Periodic Table** is not merely to predict future discoveries, but to provide a systematic methodology for identifying **where future discoveries are most likely to occur**.

Section 4. Summary of Structural Periodicity Prediction — From Structural Repetition to Future Scientific Theories

Abstract:

Scientific prediction does not arise solely from empirical statistics or technological trends. At a deeper level, it originates from **structure itself**.

The **Periodic Table of Elements** was able to predict undiscovered elements not because it "knew the future," but because it revealed the structural principles governing elemental organization. Likewise, if the **Structural Periodic Table** successfully captures the common mechanisms underlying the formation and evolution of scientific knowledge, then its predictive power likewise derives from **structural necessity rather than empirical extrapolation**.

Accordingly, the Structural Periodic Table does not merely predict future theories or disciplines; it predicts **structural positions that have yet to be occupied**. This section summarizes the fundamental principles of structural prediction and argues that structural prediction can serve as the basis for **testable scientific research programs**, rather than remaining a purely philosophical proposal.

I. Prediction Arises from Structure, Not Experience

Conventional scientific prediction is generally based on:

- statistical analysis;
- technological trends;
- extrapolation from established laws.

Structural prediction, however, rests upon a different foundation:

- whether structural closure has been achieved;
- whether hierarchical organization is complete;
- whether structural positions remain unoccupied.

Thus, empirical prediction asks: **What may happen in the future?**

Structural prediction asks: **Which structures must eventually emerge?**

II. Three Fundamental Principles of Structural Prediction

2.1 The Principle of Structural Completeness

Whenever a structural system has not yet achieved complete closure, its missing structural positions will eventually be filled.

Therefore, **structural vacancies indicate future directions of development.**

2.2 The Principle of Structural Symmetry

When multiple related disciplines exhibit common structural patterns, the asymmetrical portions tend naturally toward the formation of new unified structures.

Consequently, symmetry not only explains existing theories but also predicts future ones.

2.3 The Principle of Structural Stability

Any scientific theory capable of long-term survival must satisfy three fundamental structural roles: **Generation → Evolution → Closure**

A theory that fails to achieve structural closure over an extended period will ultimately be replaced by a more structurally complete framework.

III. Levels of Structural Prediction

Structural prediction does not directly forecast individual experimental outcomes. Instead, it operates across multiple hierarchical levels.

Level 1: Predicting new structural positions.
Level 2: Predicting new theoretical directions.
Level 3: Predicting new scientific disciplines.
Level 4: Predicting new unified scientific languages.

Accordingly, structural prediction represents a form of **Higher-Level Scientific Prediction,** rather than conventional numerical forecasting.

IV. Scientific Significance of Structural Prediction

The Periodic Table predicts: **unknown elements.**
The Structural Periodic Table predicts: **unknown structures.**
The Periodic Table organizes: **the material world.**
The Structural Periodic Table organizes: **the world of scientific knowledge.**

Its true predictive objective is therefore not future events themselves, but rather **the structural positions that future discoveries will inevitably occupy.**

V. Conclusion

Scientific progress is not simply the accumulation of knowledge. Rather, it is the continual completion of **structure**.

Whenever a structural system remains incomplete, future development acquires a definite direction. Once structural closure has been achieved, scientific development proceeds toward an entirely new hierarchical level.

Therefore, genuine scientific prediction is **not an attempt to guess the future**. It is the discovery of **positions within structure that the future must ultimately occupy.**

Concluding Statement:

The deepest form of scientific prediction is not the anticipation of future discoveries, but the identification of the structural positions that those discoveries must eventually occupy.

References (Chapter 12):

[1] **Dmitri Mendeleev** (1869). *The Relation between the Properties and Atomic Weights of the Elements*. Journal of the Russian Chemical Society.

[2] **Paul Dirac** (1928). *The Quantum Theory of the Electron*. Proceedings of the Royal Society A.

[3] **Peter Higgs** (1964). *Broken Symmetries and the Masses of Gauge Bosons*. Physical Review Letters.

[4] **Thomas S. Kuhn** (1962). *The Structure of Scientific Revolutions*. University of Chicago Press.

[5] **Herbert A. Simon** (1962). *The Architecture of Complexity*. Proceedings of the American Philosophical Society.

[6] **John Holland** (1998). *Emergence: From Chaos to Order*. Oxford University Press.

[7] **Stuart Kauffman** (1993). *The Origins of Order: Self-Organization and Selection in Evolution*. Oxford University Press.

[8] **Melanie Mitchell** (2009). *Complexity: A Guided Tour*. Oxford University Press.

[9] **Terence Tao** (2012). *Topics in Random Matrix Theory*. American Mathematical Society.

[10] **Alan Turing** (1952). *The Chemical Basis of Morphogenesis*. Philosophical Transactions of the Royal Society B.

[11] **John von Neumann** (1966). *Theory of Self-Reproducing Automata*. University of Illinois Press.

[12] **Ilya Prigogine** (1984). *Order Out of Chaos*. Bantam Books.

[13] **Chang, J.** (2026). *Structural Periodic Table: From the Chemical Periodic Table to the Origin of Life*. Universal Publishing.

[14] **Chang, J.** (2026). *Recursive Self-Organizing Evolutionary Breakthrough: Unifying Natural, Social and Life Sciences*. Universal Publishing.

[15] **Chang, J.** (2013). *Grand Ultimate Theory*. Universal Publishing.

[16] **Chang, J.** (2003-2006). *Universal Law*. Universal Publishing.

Chapter Fourteen: Theoretical Boundaries of the Structural Periodic Table and the Future of Science

Figure 3-14: A crop circle illustrating a Galactic Civilization. See book 《Crop Circle》 for details.

Chapter Overview:

The preceding chapters have established the theoretical foundation, applications, and predictive framework of the **Structural Periodic Table**.

If the Structural Periodic Table correctly captures the mechanisms underlying structural formation, it should not only explain existing structures but also predict structures that have not yet been discovered. However, the value of a mature scientific theory does not lie solely in what it can explain.

More importantly, it must also recognize **what it cannot explain**. The history of science demonstrates that every successful theory possesses a well-defined domain of applicability.

- Newtonian mechanics cannot accurately describe motion at relativistic speeds.
- Classical electromagnetism cannot fully explain quantum phenomena.
- Classical geometry cannot describe the curvature of spacetime.

Scientific progress, therefore, is not merely a continual expansion of explanatory power, but an ongoing process of discovering the limits of existing theories. The Structural Periodic Table is no exception.

If it represents a genuine structural framework of nature, then it must be subjected to the same standards applied to every scientific theory:

- Does a higher structural hierarchy exist beyond the present framework?
- Can the Structural Periodic Table explain its own foundations?
- Are there domains that remain fundamentally beyond structural explanation?
- Under what conditions should the theory be revised—or even abandoned?

Accordingly, this chapter no longer focuses on **what the Structural Periodic Table can explain**, but rather on **what it cannot explain**. The discussion is organized into four progressive stages:

Sect	Central Question	Structural Perspective
1	If the Structural Periodic Table is valid, does a Fifth Ring exist?	Upward Exploration
2	Can the Structural Periodic Table explain itself?	Self-Adjudication
3	Which phenomena may remain permanently beyond structural explanation?	Theoretical Boundaries
4	Under what conditions must the Structural Periodic Table be revised or abandoned?	Theoretical Adjudication

Table 3-14-1: The Boundary Framework of the Structural Periodic Table.

From a logical perspective, these four questions form a progressive sequence: **Expansion → Reflection → Boundary → Adjudication**

Specifically,

- **Section 1** explores how far the theory may be extended.
- **Section 2** examines whether the theory can evaluate its own foundations.
- **Section 3** investigates the domains beyond the reach of structural explanation.
- **Section 4** establishes the conditions under which the theory ceases to remain valid.

Consequently, this chapter completes the final methodological cycle of the Structural Periodic Table:

- **Establishing Structure**
- **Applying Structure**
- **Predicting Structure**
- **Adjudicating Structure**

A scientific theory demonstrates genuine maturity only when it is willing to subject itself to its own standards of evaluation.

Therefore, the ultimate concern of Chapter 14 is not the future of the Structural Periodic Table itself, but **how the Structural Periodic Table confronts the limits of its own explanatory power**.

Section 1. From Structural Explanation to Structural Generation – Does Science Require a Fifth Structural Ring?

Abstract:

The **Structural Periodic Table** is not intended as a metaphor, but as a set of generative principles.

If the **Point–Line–Circle** framework, together with the **Three-Ring–Four-Ring hierarchy**, genuinely captures the generative logic of mathematics, the organizational logic of physics, and the stabilizing logic of chemistry, then it represents more than an explanatory model—it approaches an **axiomatic system for structural generation**.

Once such an axiomatic framework demonstrates consistency across multiple scientific disciplines, it no longer functions merely as a scientific theory; rather, it becomes part of the structural foundation of science itself.

In this perspective, the **Fourth Ring** answers the question: **Which structures are permitted to exist?**

The potential **Fifth Ring**, however, addresses a deeper question: **Which structures must inevitably emerge?**

I. Introduction

Within the Structural Periodic Table, four structural layers have already been established:

- **First Ring:** Point • — Irreducible Existence
- **Second Ring:** Line | — Directional Evolution
- **Third Ring:** Circle O — Structural Closure and Stability
- **Fourth Ring:** Feedback Adjudication — Selection through Generativity and Structural Admissibility

A natural question therefore arises: Can structural organization continue beyond the Fourth Ring?

Does a **Fifth Structural Ring** exist?

This is not simply a desire to extend the framework. Rather, it is a question concerning the internal consistency and completeness of structural theory itself.

II. Conditions for the Emergence of a New Structural Ring

Within the Structural Periodic Table, a new hierarchical ring can emerge only if three conditions are simultaneously satisfied:

1) The preceding structural level has achieved complete closure.
2) Irreducible couplings arise between lower-level structures.
3) The newly emerging phenomena can no longer be fully explained within the existing Three-Ring or Four-Ring framework.

Unless all three conditions are met simultaneously, no new structural level is required.

III. What Have the First Four Rings Already Accomplished?

The progression of structural closure can be summarized as follows:

Ring	**Structural Function**
First Ring	Establishment of Local Structure
Second Ring	Formation of Evolutionary Pathways
Third Ring	Stable Structural Closure
Fourth Ring	Adjudication of Structural Admissibility

Table 3-14-1-1: Structural Closure from the First to the Fourth Ring.

At present, the Fourth Ring is responsible for:

- adjudicating structural generativity;
- selecting structurally admissible systems;
- enforcing consistency among competing theoretical models.

Up to the Fourth Ring, science primarily determines **which structures are permitted to exist.**

This remains a fundamentally passive criterion. If all questions concerning structural admissibility can be resolved within the Fourth Ring, then no logical space remains for a Fifth Ring.

IV. Under What Conditions Could a Fifth Ring Emerge?

A Fifth Structural Ring would become necessary only under one or more of the following circumstances.

4.1 Structural Self-Reorganization

If structures become capable not only of generating new systems but also of rewriting their own generative rules, a higher-order level of **self-governing structure** would emerge.

4.2 Structural Openness Beyond Existing Closure

If future observations reveal stable systems that satisfy neither the Three-Ring nor the Four-Ring criteria, yet nevertheless remain persistently stable, then the present structural framework would no longer be complete.

4.3 Cross-Scale Irreducibility

If certain phenomena cannot be reduced to:

- Point (irreducible units),
- Line (evolutionary pathways),
- Circle (structural closure),
- or Fourth-Ring adjudication,

then an additional structural hierarchy may be required.

V. What Would the Fifth Ring Change?

If a Fifth Structural Ring exists, it would no longer ask:

- How do structures exist?
- Which structures are permitted?

Instead, it would ask a fundamentally new question: **Which structures must inevitably emerge?**

Science would consequently shift from

- describing nature, and
- explaining nature, toward
- predicting structures that have not yet appeared.

VI. Expected Characteristics of the Fifth Ring

Logically, the Fifth Ring cannot simply be

- a larger Point,
- a longer Line,
- or a bigger Circle.

Instead, structural reasoning suggests three defining characteristics.

6.1 Generativity

Rather than merely judging structural existence, it would determine

- generative pathways,
- generative conditions,
- and the likelihood of structural emergence.

6.2 Self-Consistency

A Fifth-Ring theory must remain internally consistent

- across scientific disciplines,
- across hierarchical levels,
- and across multiple spatial and temporal scales.

6.3 Predictive Structural Vacancies

Just as Mendeleev predicted undiscovered chemical elements, a Fifth Ring should identify **vacant positions corresponding to theories that have not yet emerged.**

In this sense, the Fifth Ring would not merely organize structures. It would organize the **rules governing structures themselves**. It therefore represents a genuine **meta-structural level**.

VII. Is There Any Evidence for a Fifth Ring?

At present,

- mathematics exhibits structural closure within the Three-Ring framework;
- physics likewise achieves Three-Ring closure;
- chemistry demonstrates Three-Ring closure;
- and the Fourth-Ring adjudication mechanism appears sufficient.

No empirical evidence currently requires introducing a Fifth Structural Ring. Accordingly, the present Structural Periodic Table remains a **closed — but open to future revision — framework**.

VIII. A Transformation in the Nature of Science

Should a Fifth Ring eventually emerge, science itself would undergo a profound transition — from being **equation-driven** to becoming **structure-driven**.

Future scientific inquiry would no longer ask merely: *"How should this equation be solved?"*

Instead, it would increasingly ask: *"Must this structure inevitably emerge?"*

IX. A Statement of Theoretical Risk

If future discoveries demonstrate that:

- certain theories are structurally impossible;
- some research directions can never achieve structural closure;
- existing frameworks fail because of irreducible cross-level phenomena;
- or structures are capable of spontaneously rewriting their own governing principles,

then the Structural Periodic Table must itself evolve.

Otherwise, it risks becoming a closed doctrine rather than an open scientific framework. This is not criticism. It is the natural consequence of structural selection.

X. The True Meaning in the Age of Artificial Intelligence

The emergence of a Fifth Ring may coincide naturally with the development of artificial intelligence. This is because:

- AI fundamentally functions as a **structural search engine**.
- The Fifth Ring would function as a **structural generation engine**.

When structural axioms become integrated with machine-guided discovery, science may enter an era of **automated structural generation**.

XI. Key Conclusion

At present,

- the Structural Periodic Table does not require a Fifth Ring;
- nor does it exclude the possibility of one.

It therefore remains **an open yet structurally complete system.**

If a Fifth Ring eventually emerges, it will not merely extend the present framework. It will redefine what **structure itself** means.

XII. Prediction

- The First Ring enables structures to appear.
- The Second Ring enables structures to evolve.
- The Third Ring enables structures to stabilize.
- The Fourth Ring determines which structures are permitted to exist.
- The Fifth Ring — if it exists — will determine which structures must inevitably emerge.

If the Structural Periodic Table is fundamentally correct, then the Fifth Ring does not belong solely to the future. **Its emergence may already have begun.**

Section 2. Can Structural Theory Adjudicate Itself? – The Ultimate Test of the Structural Periodic Table

Abstract:

This section subjects the **Structural Periodic Table** itself to an extreme theoretical examination.

If the Structural Periodic Table is proposed as a generative framework capable of unifying multiple scientific disciplines, it must satisfy three ultimate criteria:

- **structural minimality**,
- **closure and uniqueness**, and
- **falsifiability**.

This chapter argues that the Structural Periodic Table possesses scientific legitimacy only if it is capable of adjudicating itself and explicitly accepts the possibility of being revised or refuted.

If it proves to be irreducible, unfalsifiable, or irreplaceable in principle, it ceases to function as a scientific framework and instead becomes a metaphysical assumption.

I. First Adjudication: Is It Truly the Minimum Generative Structure?

The Structural Periodic Table proposes that **Point • → Line | → Circle O,** constitutes the minimum set of structural roles required for the generation of all organized systems.

However, an essential question must be asked: Can an even simpler structural language accomplish the same task?

For example:

- Is a binary structure sufficient?
- Can the entire framework be reduced to a single recursive principle?

• Does there exist a more abstract generative language requiring fewer fundamental elements?

If a simpler framework can successfully explain mathematics, physics, chemistry, and life using fewer generative principles, then the Point–Line–Circle framework would no longer qualify as the minimum structural language.

Criterion: If a future unified framework explains the same range of scientific structures using fewer fundamental generative principles, the Structural Periodic Table loses its claim to structural minimality.

II. Second Adjudication: Is the Structural Closure Truly Complete and Unique?

The Structural Periodic Table assumes that the three structural roles, together with multiple hierarchical levels, form a finite and closed organizational system.

A further question naturally follows: Could there exist

- a fourth fundamental structural role?
- a primitive generative element independent of Point, Line, and Circle?
- an irreducible phenomenon that transcends all three structural roles?

If such an irreducible structural primitive exists, then the Structural Periodic Table represents only a subsystem rather than a complete structural framework.

Criterion: If a reproducible phenomenon cannot be consistently represented by the generative logic of Point, Line, and Circle, the completeness of the Structural Periodic Table must be reconsidered.

III. Third Adjudication: Is the Theory Falsifiable?

A genuine scientific framework must allow the possibility of its own failure. Accordingly, the Structural Periodic Table must remain open to situations in which:

- a scientific discipline develops a fundamentally new mode of structural generation beyond the existing framework;
- a verified counterexample contradicts one of its structural principles;
- or an observed phenomenon consistently violates hierarchical closure.

If the Structural Periodic Table is formulated in such a way that every possible observation can always be interpreted as supporting it, then it is no longer a scientific structure but merely an interpretive framework. A scientific theory must preserve not only explanatory power but also the possibility of refutation.

IV. Fourth Adjudication: Is It Merely a Self-Referential Encoding?

Perhaps the most difficult question is whether the Structural Periodic Table simply re-labels existing scientific knowledge.

If it can only

- rename established disciplines,
- without predicting previously unknown structures,
- and without adjudicating future theoretical developments,

then it functions merely as a mapping system rather than a genuinely generative framework.

A structural theory should generate new knowledge, not simply reorganize existing knowledge.

V. Fifth Adjudication: Can It Be Independently Verified by Artificial Intelligence?

If the Structural Periodic Table represents an objective structural framework, then its logical organization should not depend upon its original author.

An artificial intelligence system should, in principle,

- reconstruct its internal logic solely from its formal rules,
- evaluate structural consistency,
- and reproduce its conclusions independently.

If the framework can only be understood through continual human interpretation, then its formalization remains incomplete.

Independent reconstruction therefore becomes an important future criterion for evaluating the maturity of structural theory.

VI. Ultimate Conclusion

The Structural Periodic Table can remain scientifically valid only if it continuously accepts the possibility of being replaced by a simpler framework. It must allow irreducible counterexamples. It must remain open to revision.

Otherwise, it risks becoming precisely the type of universal explanatory framework that it seeks to avoid—a theory capable of explaining everything, and therefore explaining nothing.

VII. Summary

A mature structural theory must ultimately be capable of demonstrating the conditions under which it becomes unnecessary.

If one day a simpler generative framework succeeds in unifying mathematics, physics, chemistry, life science, and other disciplines more economically than the Structural Periodic Table, then the Structural Periodic Table should willingly yield its place.

The purpose of structural theory is not to exist forever, but to identify the simplest principles capable of generating the greatest diversity of stable structures.

Section 3. The Boundaries of Structural Explanation — Which Phenomena May Never Be Fully Described by Structure?

Abstract:

The **Structural Periodic Table** aims to characterize the organizational principles governing scientific systems across multiple hierarchical levels. Through the **Point–Line–Circle** framework and its multi-ring architecture, it seeks to explain three fundamental structural properties: **irreducibility**, **directional evolution**, and **stable structural closure**.

However, no structural theory can remain scientifically meaningful unless it clearly defines its own limits. Without explicit boundaries, any explanatory framework risks becoming overly generalized or drifting into metaphysical speculation.

Accordingly, this section argues that the **Structural Periodic Table is not intended to explain every phenomenon**, but rather a specific class of structural phenomena.

I. What Can the Structural Periodic Table Explain?

The Structural Periodic Table addresses questions concerning structural organization, including:

- whether a structure is capable of existing,
- whether its evolution can be sustained,
- whether stable closure can be achieved,
- and whether different theoretical descriptions converge toward a common structural framework.

In other words, the Structural Periodic Table investigates **structural admissibility**, rather than every causal detail of natural phenomena.

II. Three Classes of Phenomena Beyond the Scope of Structural Explanation

2.1 Initial Conditions Governed by Contingency

Structural theory can determine

- which configurations are capable of stable existence,
- and which evolutionary pathways are structurally admissible.

However, it cannot explain:

- why the universe began from one particular initial condition rather than another,
- the precise microscopic outcome of an individual quantum fluctuation,
- or the accidental triggering of a specific historical event.

Initial conditions precede structural organization itself.
Structural theory evaluates what is structurally permitted; it does not determine historical contingency.

2.2 Subjective Experience and First-Person Consciousness

Structural theory may explain

- how psychological organization emerges,
- and why certain cognitive or social structures remain stable.

However, it does not explain:

- why a particular individual experiences a specific emotion at a particular moment,
- or the qualitative character of conscious experience itself.

The first-person aspect of consciousness lies outside the scope of structural adjudication. Structural theory describes systems rather than subjective experience.

2.3 Technical and Computational Details

The Structural Periodic Table does not replace:

- precise mathematical calculations,
- numerical simulations,
- solutions of specific differential equations,
- or experimental error analysis.

It is neither

- a computational tool,
- a numerical model,
- nor an engineering algorithm.

Instead, it provides a framework for evaluating **structural organization and admissibility.**

III. The Most Serious Misuses of Structural Theory

The following interpretations would fundamentally distort the purpose of the Structural Periodic Table.

3.1 Confusing Structural Principles with Detailed Causal Mechanisms

Structural theory explains **why certain organizational forms are possible**, not every intermediate causal process through which they emerge.

3.2 Treating Structural Similarity as Literal Identity

Structural correspondence does not imply conceptual equivalence. For example, similarities between mass concentration and political centralization do **not** imply that the universe possesses political intentions.

Structural analogy represents mathematical or organizational correspondence rather than anthropomorphic interpretation.

3.3 Treating Structural Theory as Ultimate Truth

Even if the Structural Periodic Table proves successful, it should still be regarded as a theory describing **organizational principles**, rather than an ultimate explanation of reality itself.

Scientific theories explain particular aspects of nature; none should be mistaken for absolute truth.

IV. The Fundamental Boundary of the Structural Periodic Table

Its scope may be summarized in a single statement: The Structural Periodic Table evaluates whether structures are capable of existing and remaining stable, but it does not replace historical contingency, subjective experience, or detailed computational description.

The Structural Periodic Table must be revised — or, if necessary, abandoned — should future evidence demonstrate any of the following:

- a stable system that cannot be represented within the Point–Line–Circle framework,
- a physically verified theory whose stable organization contradicts the three structural principles,
- or a complex system whose persistent stability cannot be explained through hierarchical structural evolution.

Such discoveries would indicate that the present framework is incomplete and requires extension or replacement.

V. Final Boundary Statement

A scientific theory becomes stronger not by claiming to explain everything, but by clearly identifying what lies beyond its explanatory reach. The strength of the Structural Periodic Table therefore lies not only in what it explains, but also in the boundaries it openly acknowledges.

Section 4. Under What Conditions Must the Structural Periodic Table Be Revised or Abandoned?

Abstract:

A scientific theory cannot truly qualify as science unless it explicitly defines the conditions under which it may fail.

The **Structural Periodic Table** does not claim to be an eternally valid framework. On the contrary, it must remain subject to continual examination through future experiments, mathematical developments, and alternative theoretical frameworks.

Should future discoveries reveal stable, reproducible, and empirically verified phenomena that cannot be incorporated within the framework of Structural Periodicity, then the theory must be revised — or, if necessary, replaced by a more comprehensive framework.

Accordingly, this section is concerned not with why the Structural Periodic Table may be correct, but with the circumstances under which it must acknowledge its own limitations. Defining such conditions represents not only the boundary of the theory itself, but also one of the essential principles of scientific inquiry.

I. Scientific Progress Is Primarily a Process of Revision Rather Than Rejection

The history of science demonstrates that successful theories are rarely discarded entirely. More commonly, they are revised under broader conditions and become limiting cases within more general theories. For example:

- **Newtonian mechanics** successfully explained low-speed macroscopic motion, yet required modification through relativity under conditions of high velocity and strong gravitation.
- **Maxwell's electromagnetic theory** unified electricity and magnetism but could not account for quantum phenomena, leading to the development of quantum theory.

• **Einstein's theory of relativity** established the modern geometric description of spacetime, yet has not achieved complete unification with quantum theory.

Scientific progress therefore does not consist of continuously overthrowing previous theories, but rather of extending their domains of applicability while clarifying their limitations.

The Structural Periodic Table must be judged according to the same scientific standard.

Should a future theory provide a broader and more powerful explanation of nature, the Structural Periodic Table should likewise become a special case within that more comprehensive framework.

II. First Condition: The Discovery of Stable Systems Without Structural Organization

The central hypothesis of the Structural Periodic Table is that every system capable of long-term stability possesses some form of structural generation, evolution, and closure.

Suppose future research identifies a system that:

- remains stably organized over long periods,
- is repeatedly reproducible,
- performs complex functions,
- yet possesses neither identifiable generative units (Point),
- nor evolutionary pathways (Line),
- nor stable structural closure (Circle),

and furthermore cannot be embedded within any higher-order structural hierarchy.

Such evidence would imply that structural organization is not a necessary condition for stable existence.

Under these circumstances, one of the fundamental assumptions of Structural Periodicity would require revision.

It should be emphasized that this criterion concerns experimentally verified natural phenomena rather than speculative hypotheses.

III. Second Condition: Independent Theories Never Converge

The Structural Periodic Table assumes that theories describing different scientific domains should ultimately converge toward a higher structural unity.

If future evidence demonstrates that:

- multiple theories remain simultaneously valid,
- yet can never be unified,
- and no higher-level framework exists capable of reconciling them,

then the hypothesis of structural convergence would no longer hold. In that case, the Structural Periodic Table could continue to function as a descriptive framework, but could no longer claim to provide a unified theory.

IV. Third Condition: Stable Systems Without Hierarchical Organization

This book argues that natural systems evolve through successive hierarchical transitions. If future research establishes the existence of systems that possess:

- no hierarchical organization,
- no scale-dependent structure,
- and no structural recursion,

yet nevertheless remain permanently stable and reproducible, then hierarchical organization cannot be regarded as a necessary condition for stability.

Accordingly, the hierarchical component of Structural Periodicity would require substantial revision.

V. Fourth Condition: Persistent Failure of Predictive Capability

A scientifically meaningful theory should not only explain existing observations but also possess predictive capability.

Accordingly, if the Structural Periodic Table consistently predicts that certain structures cannot exist, while future experimental evidence repeatedly demonstrates that these structures:

- exist stably,
- are reproducible,
- and cannot be incorporated into the Structural Periodicity framework,

then the theory must:

- revise its fundamental axioms,
- modify its predictive principles,
- or acknowledge that its predictive mechanism has failed.

Predictive failure remains one of the strongest criteria by which any scientific theory is evaluated.

VI. Fifth Condition: Loss of Independence of the Structural Evolution Equation

This book proposes

$$dM/dt = \alpha_1 \nabla M + \alpha_2 I(E,S,C) + \alpha_3 Q(x)$$

as a unified expression of structural generation.

However, if future research rigorously demonstrates that this equation:

- is merely a reformulation of existing mathematical models,
- provides no genuinely new explanatory capability,
- and yields no novel predictive power,

then it can no longer serve as the central equation of the Structural Periodic Table. A new structural generative framework would then need to be established.

VII. Sixth Condition: The Emergence of an Alternative Structural Language

The Structural Periodic Table proposes that **Point •, Line |, and Circle O** constitute the minimum structural language shared across the natural sciences.

Nevertheless, if an entirely different structural language were developed that:

- does not depend upon Point, Line, or Circle,
- successfully explains the full range of natural phenomena,
- and possesses equal or greater explanatory and predictive power,

then Point–Line–Circle would no longer represent a unique structural foundation. The Structural Periodic Table would therefore need to acknowledge the new framework and reinterpret its own role within it.

For this reason, this book does **not** claim that Point, Line, and Circle constitute the ultimate language of the universe. Rather, they are proposed as the simplest, most unified, and currently most explanatory structural language available.

VIII. The Greatest Risk Is Not Opposition

Every new scientific theory encounters criticism.

The genuine danger is not disagreement, but the possibility that the Structural Periodic Table may ultimately prove to be only an elegant metaphor rather than a genuine scientific framework.

If it cannot clearly determine:

- which structures are admissible,
- which structures are impossible,
- and which predictions are experimentally testable,

then it remains philosophy rather than science.

IX. Abandonment Does Not Mean Failure

Should the Structural Periodic Table eventually require revision — or even replacement — this should not be regarded as failure. Instead, it would indicate that science has discovered an even higher principle of organization.

Newton did not fail. Einstein did not fail. Both became special cases within broader theoretical frameworks.

Likewise, a mature scientific theory is not one that remains permanently correct, but one that recognizes the circumstances under which it should willingly yield to a more comprehensive theory.

X. Summary

The Structural Periodic Table should be revised — or, if necessary, abandoned — only if future research establishes a stable, reproducible, and universally consistent mode of natural organization that:

- cannot be represented within any hierarchical structural framework,
- cannot be explained through structural generation, structural evolution, structural closure, or structural admissibility,
- and is simultaneously supported by experimental evidence, mathematical proof, and independent theoretical analysis.

Under such circumstances, the Structural Periodic Table should give way to a more comprehensive unified theory.

Concluding Statement:

A scientific theory reaches maturity not because it claims to explain everything, but because it clearly defines the conditions under which it must be revised — or gracefully leave the stage of scientific history.

Section 5. Concluding Remarks on the Methodology of Future Science — From Equation-Based Science to Structure-Guided Science

Abstract:

The evolution of science is reflected not only in the accumulation of knowledge, but also in the continuous evolution of scientific methodology. From empirical observation to mathematical equations, from unified theories to artificial intelligence-assisted discovery, every major scientific revolution has been accompanied by the emergence of a new scientific language.

This book proposes that, if the **Structural Periodic Table** can serve as a unified cross-disciplinary structural framework, its greatest significance lies not merely in introducing another scientific theory, but in suggesting a new scientific methodology—**Structure-guided Science**.

Rather than replacing mathematics, physics, or chemistry, this methodology seeks to establish a common structural language across scientific disciplines. It encourages scientific inquiry to evolve gradually from **solving equations** toward **discovering structures**, and ultimately toward a new era in which artificial intelligence assists the discovery of previously unknown structural principles.

I. The Historical Evolution of Scientific Methodology

A review of the history of science reveals several major stages in the evolution of scientific language.

1.1 Stage One: Empirical Science

The earliest stage of science was primarily based upon observation and experiment. Scientific inquiry sought to answer questions such as:

- What is the world composed of?
- How does nature appear?

1.2 Stage Two: Equation-Based Science

With the development of mathematics, scientific knowledge increasingly became expressed through equations.

Representative achievements include:

- Newton's equations of motion;
- Maxwell's equations;
- Schrödinger's equation;
- Einstein's field equations.

During this stage, science focused primarily on answering one central question: **How does nature evolve?**

1.3 Stage Three: Unified Theories

Modern science has increasingly pursued the unification of apparently independent theories. Representative examples include:

- Electroweak Unification;
- Grand Unified Theories;
- String Theory;
- M-Theory.

The central question became: **Do different physical theories originate from a common underlying framework?**

1.4 Stage Four: AI-Assisted Science

Recent advances in artificial intelligence have begun transforming the process of scientific discovery.

AI systems can now:

- analyze large datasets automatically;
- search theoretical model spaces;
- generate scientific hypotheses;
- and increasingly participate in the discovery of new scientific principles.

Science is therefore entering an era of **human–AI collaborative discovery**.

II. Future Science Requires a New Common Language

As scientific disciplines continue to become increasingly specialized, communication between different fields becomes progressively more difficult.

Mathematics possesses its own formal language. Physics possesses another. Chemistry, life science, information science, and the social sciences each continue to develop increasingly specialized conceptual systems.

Meanwhile, many of today's most important scientific problems extend across traditional disciplinary boundaries.

Future science therefore requires not merely additional knowledge, but a **shared structural language** capable of connecting multiple disciplines.

III. The Structural Periodic Table as a Structural Language

If the Structural Periodic Table proves to be valid, it should be understood not as another specialized body of scientific knowledge, but as a new descriptive framework. Its primary concerns are:

- how structures are generated;
- how structures evolve;
- how structures achieve stability;
- and how structures are evaluated through structural adjudication.

Accordingly, its subject is not confined to any single discipline, but instead concerns organizational principles shared across many scientific domains.

Mathematics, physics, and chemistry may therefore be viewed as different manifestations of the same underlying structural language operating at different hierarchical levels.

In this sense, the Structural Periodic Table is closer to a **scientific grammar** than to an additional branch of natural science.

IV. Structural Science in the Age of Artificial Intelligence

Future scientific discovery will likely depend not only upon human intuition but increasingly upon artificial intelligence exploring theoretical spaces.

However, AI ultimately requires more than large quantities of data. It requires efficient principles for searching.

Without a unified structural language, AI can only search through enormous collections of equations.

With such a language, AI may instead search for:

- structural vacancies;
- theoretical gaps;
- new generative pathways;
- previously unknown stable organizational patterns.

Consequently, future AI systems may discover not merely new mathematical formulas, but entirely new structural relationships.

Structure-guided Science may therefore become an important bridge between artificial intelligence and scientific discovery.

V. A Possible New Paradigm for Future Science

The evolution of scientific methodology may gradually follow the sequence: **Empirical Observation → Mathematical Description → Unified Theory → AI-Assisted Discovery → Structure-guided Science**

Within this emerging paradigm, the central scientific question gradually shifts from: **"How should this equation be solved?"** to **"Is this structure capable of existing?"**

Scientific progress may increasingly emphasize structural organization rather than computational complexity.

VI. An Open Future

This book does not claim that the Structural Periodic Table represents a completed theory. On the contrary, it should be regarded as an open framework. Future developments may introduce:

- simpler generative languages;
- higher-order structural principles;
- additional structural roles;
- or even more general unified frameworks.

Should a future theory explain a broader range of phenomena using fewer fundamental principles, the Structural Periodic Table should itself be revised — or replaced.

Such continual refinement represents the normal progression of science.

VII. Final Summary

The **Periodic Table of Elements** organized the material world.

The **Structural Periodic Table** seeks to organize the structural relationships underlying knowledge itself.

Its purpose is not to replace existing scientific disciplines, but to provide a common structural language through which different sciences may be compared, connected, evaluated, and — when necessary — revised.

If future science gradually evolves from being **equation-driven** toward being **structure-guided**, then the greatest contribution of the Structural Periodic Table may ultimately lie not in the theory itself, but in offering a possible direction for the future development of scientific thought.

Concluding Statement: Scientific progress is driven not only by new discoveries, but also by new languages. The Structural Periodic Table does not seek to establish yet another scientific discipline; rather, it explores the possibility of a common structural language capable of connecting all scientific disciplines.

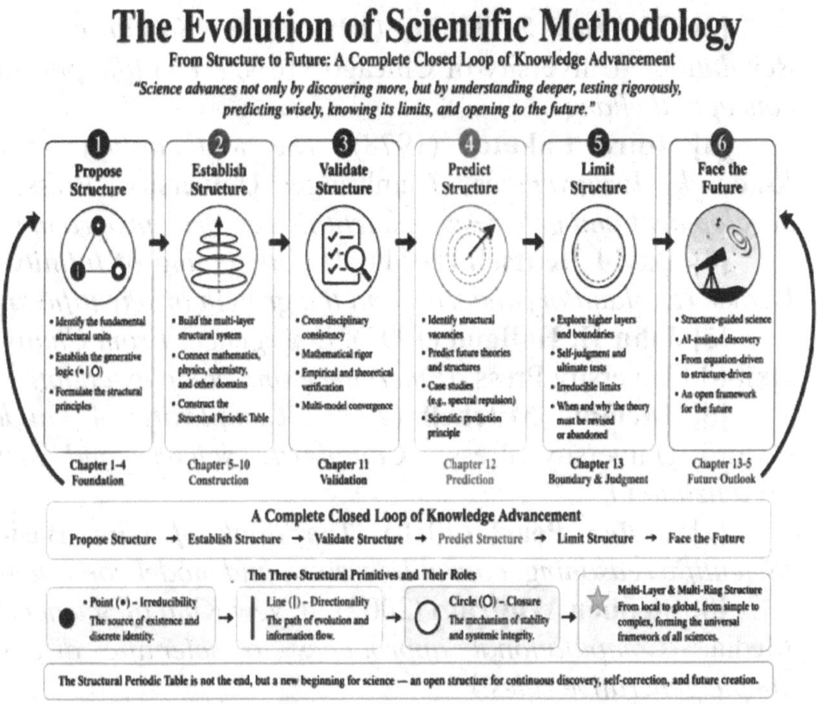

Figure 3-14-5-1: The Evolution of Scientific Methodology.

Caption: The logical development of the **Structural Periodic Table** forms a complete methodological cycle:

Proposing Structures → Establishing Structures → Verifying Structures → Predicting Structures → Defining Boundaries → Exploring the Future.

If future science ultimately requires a unified structural language, then the **Structural Periodic Table** may be regarded as one exploration of that possibility.

References : (Chapter 14)

[1] Karl Popper (1959). *The Logic of Scientific Discovery.* Routledge. *(Scientific falsifiability and the demarcation of science.)*

[2] Thomas S. Kuhn (1962). *The Structure of Scientific Revolutions*. University of Chicago Press. *(Scientific paradigms and conceptual change.)*

[3] Imre Lakatos (1978). *The Methodology of Scientific Research Programmes*. Cambridge University Press. *(Theory development and progressive scientific research programmes.)*

[4] David Deutsch (2011). *The Beginning of Infinity*. Penguin Books. *(Explanatory theories and the growth of scientific knowledge.)*

[5] John H. Holland (1998). *Emergence: From Chaos to Order*. Oxford University Press. *(Emergence and complex adaptive systems.)*

[6] Melanie Mitchell (2009). *Complexity: A Guided Tour*. Oxford University Press. *(Complexity science and hierarchical organization.)*

[7] Judea Pearl (2018). *The Book of Why*. Basic Books. *(Scientific reasoning, causal inference, and model construction.)*

[8] Stephen Wolfram (2002). *A New Kind of Science*. Wolfram Media. *(Computational approaches to scientific discovery and simple generative rules.)*

[9] Herbert A. Simon (1996). *The Sciences of the Artificial* (3rd ed.). MIT Press. *(Hierarchy, complexity, and artificial systems.)*

[10] Murray Gell-Mann (1994). *The Quark and the Jaguar: Adventures in the Simple and the Complex*. W. H. Freeman. *(Simple rules, complexity, and emergence.)*

[11] Chang, J. (2026). *Recursive Self-Organizing Evolutionary Breakthrough — Unifying Natural, Social and Life Sciences*. Universal Publishing.

[12] Chang, J. (2013). *Grand Ultimate Theory*. Universal Publishing.

[13] John von Neumann (1966). *Theory of Self-Reproducing Automata*. University of Illinois Press. *(Self-organization, recursive generation, and theoretical foundations of complex systems.)*

Part IV – Epilogue

From the Periodic Table of Elements to the Structural Periodic Table – A Reflection Beyond the Book

Figure 3-15: A crop circle depicting a Galactic Civilization. For details, see book 《Crop Circle》.

I. The Periodic Table of Elements and the Structural Periodic Table

The **Periodic Table of Elements** transformed chemistry. Its greatness lies not merely in organizing more than one hundred chemical elements, but in revealing an underlying order hidden within nature: seemingly different substances can be organized according to a unified structural principle.

For more than a century, the periodic table has continued to evolve, yet it has always addressed essentially the same question: **How is matter organized?**

This book attempts to ask a different question: **If matter exhibits structural periodicity, might knowledge itself also possess structural periodicity?**

Accordingly, this book is concerned not simply with mathematics, physics, or chemistry as individual disciplines, but with the possibility that these sciences share a deeper structural language.

From elements, reactions, and molecular configurations to living systems; From mathematics, physics, and chemistry to theories that may emerge in the future; the central objective of this work has remained unchanged: **to search for a common structural organization underlying different scientific disciplines.**

II. The objective of the Structural Periodic Table

If such a common structure exists, then the **Structural Periodic Table** is only one possible expression of it. If future research discovers a simpler, more unified structural language, this book should likewise be revised — or even replaced. For science has never sought to preserve any particular theory forever.

Its true purpose is to move continually toward deeper structural principles. The Structural Periodic Table proposed in this volume may be correct. It may also be incomplete. It may represent only an initial framework. Or perhaps merely a stepping stone toward a more mature theory still to come.

Regardless of its ultimate fate, it attempts to address a question that has accompanied scientific inquiry throughout history: **Do different scientific disciplines share a common structural language?**

If future research ultimately demonstrates that the answer is negative, then this book will nevertheless have fulfilled its purpose as an exploration. If future research confirms even part of its central ideas, then what matters most will not be this particular book, but the question it has helped to formulate.

Perhaps scientific progress results not only from discovering new facts, but also from discovering new ways of organizing knowledge. Perhaps the future unification of science will begin not with another equation, but with another language — a language of structure.

By the completion of this volume, the overall framework has naturally evolved into four interconnected parts:

- **Part I — Establishing Structure:** *What is structure?*
- **Part II — Structural Foundations in the Natural Sciences:** *Why does the framework hold?*

• **Part III — Structural Language:** *How can structure explain different scientific disciplines?*

• **Part IV — Prediction and Future Science:** *What new possibilities does the framework suggest?*

Together, these four parts form a complete theoretical progression:

What is it? → Why does it work? → How can it be applied? → Where might it lead?

This progression closely follows the standard architecture of a mature theoretical framework.

Consequently, the **Structural Periodic Table** establishes a coherent conceptual pathway: **Establish Structure → Verify Structure → Develop a Structural Language → Predict the Unknown → Examine the Theory's Own Boundaries**

III. 《Structural Periodic Table》& 《Hexagonal Structures》

If the forthcoming volume, **Hexagonal Structures**, continues this line of development, the relationship between the two books becomes naturally complementary.

The **Structural Periodic Table** seeks to answer: **Why does nature exhibit recurring structural organization?**

Hexagonal Structures will seek to answer: **How do these structures further evolve into the unified cycles of life, society, civilization, and the universe?**

In this way, the two works remain independently complete while forming successive stages of a broader theoretical program.

This volume explores how life may emerge from structural organization.

The next volume, **Hexagonal Structures**, will further investigate how structural organization develops into animals, plants, psychology, politics, economics, and ultimately civilization itself, forming higher-order cycles of living systems.

Final Statement: The Periodic Table of Elements organized the material world. The Structural Periodic Table attempts to organize the world of knowledge. Whether or not this particular attempt ultimately succeeds, the search for a common structural language will continue to accompany the future development of science.

Each generation of science inherits the knowledge of its predecessors while searching for its own language for understanding nature. Perhaps the next language will be the language of structure.

IV. Back cover summary:

Structural Periodic Table explores a fundamental question: *If the Periodic Table organizes the material world, can there also exist a structural periodic table that organizes scientific knowledge?*

Beginning with mathematics, physics, chemistry, and the emergence of life, this book develops a unified structural framework based on three fundamental structural roles—**Point • (Generation), Line | (Evolution), and Circle O (Closure)**. It argues that recurring structural organization, rather than simple periodic repetition, provides a common language underlying many natural systems.

Beyond proposing a new structural framework, the book examines its predictive potential, methodological significance, and theoretical boundaries. Rather than replacing established scientific theories, it seeks to complement them by explaining why different disciplines often exhibit remarkably similar structural patterns.

Whether or not every proposal in this volume ultimately proves correct, its central aim is to stimulate a broader discussion: **Can structure itself become a universal language connecting the natural sciences?**

Key words:

- Structural Periodicity
- Structural Dynamics
- Theory of Structure
- Structural Evolution
- Unified Science
- Complex Systems

- Scientific Methodology

Other references:

[1] **John Chang**: 《Universal Law》, Nov. 2003

[2] **John Chang**: 《Great Ultimate Theory》, Sep. 2013

[3] **John Chang** : 《Crop Circle》, Oct.2015

[4] **John Chang**: 《Eastern Galactic Civilization---Unifying Religion, Philosophy and Science》, Jul. 2024

[5] **John Chang**: 《Recursive Self- Organizing Evolutionary Breakthrough----Unifying Natural, Social and Life Sciences》, Aug. 2025

[6] **John Chang**: 《Life Intelligence Wave ------ Unifying Mathematics, Physics and Chemistry》, Dec. 2025

[7] **John Chang**: 《 Metastructural Unification (Mathematics Volume): Unifying Algebra, Geometry and Probability 》, Feb. 2026

[8] **John Chang**: 《Metastructural Unification (Physics Volume) ---- Unifying Electromagnetism, Gravitation and Quantum Theory》, Apr. 2026

[9] **John Chang**: 《Metastructural Unification (Chemistry Volume): Unifying Reactions, Compounds and Elements》, May. 2026

www.ingramcontent.com/pod-product-compliance
Lightning Source LLC
LaVergne TN
LVHW040514210926
848480LV00005BA/1451